GLOBAL GEOPOLITICS

A CRITICAL

INTRODUCTION

Visit the *Global Geopolitics* Companion Website at
www.routledge.com/9780273686095 to find valuable **student**
learning material including:

- Chapter summaries and key issues
- Additional weblinks per chapter
- Discussion activities per chapter
- Geopolitics interest points

GLOBAL GEOPOLITICS

A CRITICAL

INTRODUCTION

KLAUS DODDS

Routledge
Taylor & Francis Group

LONDON AND NEW YORK

First published 2005 by Pearson Education Limited

Published 2013 by Routledge
2 Park Square, Milton Park, Abingdon, Oxon OX14 4RN
711 Third Avenue, New York, NY 10017, USA

Routledge is an imprint of the Taylor & Francis Group, an informa business

ISBN 978-0-273-68609-5 (pbk)

British Library Cataloguing-in-Publication Data
A catalogue record for this book is available from the British Library

Library of Congress Cataloging-in-Publication Data
A catalog record for this book is available from the Library of Congress

Typeset by 35 in 10.5/12.5pt Ehrhardt

For Carolyn

CONTENTS

LIST OF FIGURES

LIST OF TABLES

Supporting resources

Visit www.routledge.com/9780273686095 to find valuable online resources

Companion Website for students
- Chapter summaries and key issues
- Additional weblinks per chapter
- Discussion activities per chapter
- Geopolitics interest points

PREFACE

The audacious 11 September 2001 (hereafter September 11th) attacks on the World Trade Center (WTC) in New York and the Pentagon in Washington created a sense of moral outrage and public panic in the United States and the wider world. Nineteen men, armed with only box cutter knives, mobile phones and some flight-school training on jets in Florida, hijacked four US commercial aeroplanes and deliberately flew them into those buildings, with only the fourth missing its target (possibly the White House). About 3,000 people, of all races, religious affiliations and nationalities, perished in the attacks and few who watched the television reporting at the time will ever forget the harrowing scenes of people jumping to their deaths from the crumbling edifice of the WTC (Smith 2001). American-owned flight and telecommunications technologies were deployed as a weapon of mass destruction. A week later, the front cover of the *Economist* magazine (15–21 September 2001) announced 'The Week the World Changed.'

The September 11th attacks by a group of self-confessed Islamic militants were for some a carefully planned act of historical revenge (see Robinson 2001). Over three hundred years earlier, on 12 September 1683, the armies of the Ottoman Empire reached the gates of the Christian capital of the Hapsburg Empire, Vienna. Was the timing, therefore, a deliberate attempt to remind the world that the long-standing cultural, economic, social, political and intellectual encounter between the Christian and Islamic worlds was about to take a new turn? Perhaps high-profile American authors such as Samuel Huntington were correct in assuming that representatives of the Arab and Islamic world were intent on initiating a new 'clash of civilization' with the West? Others have pointed to the unsettled Palestine question but also to a broader process by which only Turkey, Iran, the Yemen and parts of Saudi Arabia had avoided the grip of Western imperial powers in the period between the 1683 siege of Vienna and the September 11th attacks (see Lewis 2004). While President Bush initially called for a 'crusade' against the proponents of Islamic terror, the late Palestinian-American commentator Edward Said (2001) warned of a 'clash of ignorance'.

For those operating within a shorter historical time frame, the financial and geopolitical 'heart' of the United States and the international financial system provided unwelcome evidence of a world spinning out of control. The very first thing President George W. Bush ordered, after hearing the news of those attacks, was the closure of all airports and the borders with Canada and Mexico. Within a few weeks, 600,000 jobs were lost in the United States alone as the airline industry 'downsized'. While Bush created a new Department of Homeland Security, the British prime minister, Tony Blair, ordered the creation of a new Ministerial Sub-Committee on International Terrorism. The United States and Britain, along with many other states, urgently sought to develop new policies and strategies to confront 'international terrorism' (Booth and Dunne 2002). In the process, older Cold War divisions were overturned as the United States co-operated closely with Russia in the pursuit of terror organizations around the world.

Despite the shock of September 11th, successive American governments have been pursuing proactive military and diplomatic policies in the aftermath of the Cold War. The United States, for example, now spends more on defence ($300 billion 2001–2 financial year) than any other country. American armed-forces personnel, despite the ending of the Cold War, remain stationed all over the world, including Bermuda, Columbia, Thailand, Iceland, Bosnia, Germany, Philippines, Australia, Turkey, British Indian Ocean Territory (Diego Garcia) and even Antarctica. Over 100,000 American troops are currently stationed in Europe and 30,000 in South Korea. When President Bush declared a 'war on terror' in 2001, it supplemented and extended pre-existing operations related to the 'war on drugs' in Latin America, peacekeeping in South East Europe, intelligence operations and the supply of well-established garrisons all over the world.

While post-Cold War US administrations have promoted American geopolitical and economic interests, they stand accused by others of pursuing those objectives in a unilateral (and internationally unpopular) manner. Critics contend that the United States has rejected the Kyoto climate change protocol, allowed the 1972 Anti-Ballistic Missile Treaty to lapse, refused to ratify the provisions for the International Criminal Court, imposed quotas on steel imports from the European Union and elsewhere, increased subsidies to domestic farming producers, and branded other countries such as Iran, Iraq and North Korea as part of the 'axis of evil' or simply 'rogue states'. The assault on Iraq in March 2003 was also claimed to be illustrative of unilateral behaviour as US–UK forces proceeded without the explicit authorization of the United Nations Security Council. Simultaneously, US officials and administrators have demanded that the United Nations and the international community should work together to confront new security challenges such as global terrorism and global proliferation of weapons of mass destruction. For the Bush administration, this accusation of unilateral behaviour is misplaced in the sense that America has to take steps to respond not only to the September 11th attacks but also to the possibility of future attacks. Others, however, contend that the world's only hyper-power deserves the sobriquets 'the empire of evil' and/or 'the Great Satan'.

These are uncertain, even dangerous times for all of us and *Global Geopolitics* proposes to chart how these events and others are reconfiguring the geographies of global politics and our understandings of territory, place and space. The September 11th attacks on the United States demonstrated how time-space compression brought together the terrorist and the capitalist worlds. If US Secretary of State General Colin Powell thinks that 'terrorism is the dark side of globalization' then we need to explore not only what terms like 'terrorism' might imply (given as some claim the long history of US state-sponsored terrorism in Latin America and South East Asia; see Chapter 9) but also how globalization creates interconnections and flows between people, commodities, technologies, places and regions (Urry 2002: 57). Many people in the United States believe (alongside General Powell) that cities like New York and Los Angeles are arguably more at risk than London, Paris and Berlin in the post-September 11th era. The Madrid railway bombings on 11 March 2004 with the cost of over 200 lives, however, highlighted once again European vulnerabilities to indiscriminate terror and led to the electoral defeat of the incumbent government in Spain. But where do other cities such as Baghdad, Buenos Aires and Casablanca fit into this worldview of threat perception?

Arguably, we stand on the edge of a wider historical and geopolitical cusp: as nation-states seek to maintain a rules-based international order so events and processes associated with globalization violate the territorial and military jurisdictions of states and the authority of international organizations such as the United Nations. As this happens, geopolitics as a mode of representing global political space is called into question. The traditional concern for spatial blocs, regional and global ordering and territorial presence resides uneasily in a world increasingly defined by risk, speed, hybridity and telemetricality (O Tuathail 2000). *Global Geopolitics* seeks to provide students with a sceptical audit (and a very modest one at that, given, for example, the Eurocentric view of the author) of contemporary global politics and an interpretation of how territory, space and place are redefined in these uncertain and unstable times.

Structure of the book

Chapter 1 explores contemporary examples of exclusively American geopolitical theorizing and links these discussions to the enduring uncertainties of the post-September 11th era. The intent is thus to explore how global political space is not only represented but also contested. This matters because foreign policies, national security programmes and political speeches utilize geographical descriptions such as the 'axis of evil' to legitimate and justify political practice. Chapter 2 investigates in more detail the main subject matter of this book – geopolitics and its relationship to globalization – but it does not pursue a detailed intellectual and/or historical evaluation of geopolitics. Chapter 3 shifts the focus from the Euro-American world to the manner in which contemporary geopolitics is being shaped by the interaction between North and South. While a great deal of attention is devoted to the political, economic and cultural costs of very recent events, it is essential to recall that far more children die annually from an absence of clean drinking water in the South than there are victims of terror attacks. Chapter 4 considers the role of popular culture and the ways in which television, films and other media constrain or enable particular geographical representations of the world.

Chapters 5 to 9 concentrate on some of the key issues (with associated policy challenges for governments and political leaders) shaping the contemporary global geopolitical conditions. These include the existence and proliferation of weapons of mass destruction (WMD) with a particular focus on nuclear weapons (Chapter 5), environmental degradation (Chapter 6), humanitarian intervention and human-rights protection (Chapter 7), anti-geopolitics and dissent (Chapter 8) and terror in the post-September 11th era. The concluding chapter aims to outline some of the outstanding challenges facing those seeking to understand the contemporary global geopolitical condition. Each chapter contains some additional 'In Focus' material allied with occasional definitional boxes, where I have tried to provide a little more detail and/or insight. A glossary at the end of the book provides further definitions for some of the key terms in the text. Terms which are included in the Glossary appear in bold type in the text.

While I have attempted to bring a highly contemporary focus to the book, I decided not to consider in any detail the implications of other recent events such as the capture of Saddam Hussein by American forces in Iraq. This will have to wait, along with other issues, for another edition!

Klaus Dodds, April 2004

ACKNOWLEDGEMENTS

I would like to thank Professor John Agnew, Professor Simon Dalby, Dr Jim Glassman, Dr Merje Kuus, Dr Shannon O'Lear and Dr James Sidaway for their invaluable comments and suggestions. Special thanks to Dr Anthony Budd, Dr Noel Castree, Dr Caroline Nagel, Professor Francis Robinson and Professor David Simon for their very helpful and constructively critical comments on a number of draft chapters. Jenny Kynaston was kindly responsible for producing most of the maps. Successive generations of students at Royal Holloway, University of London in combination with the generous reviews of *Geopolitics in a Changing World* have given me the confidence to produce an expanded version of this book. My colleagues at the Department of Geography at Royal Holloway, University of London continue to be immensely supportive and I am grateful to them all. None of the aforementioned bears any responsibility for any inaccuracies and/or shortcomings.

Andrew Taylor, Mary Lince, Alison Kelly, Elizabeth Harrison and the editorial team at Pearson Education were a pleasure to work with and unquestionably helped to smooth the transition from manuscript to textbook.

I am indebted once again to my mother for her advice, editing skills and intellectual guidance. Luckily for me, she also operates one of the best press clippings services in the world.

This book is dedicated to my wife, Carolyn, who remains a constant source of love and inspiration. During the writing of this book, she also gave birth to our first child, Alexander Alois Dodds.

Klaus Dodds
Royal Holloway, University of London

PUBLISHER'S
ACKNOWLEDGEMENTS

We are grateful to the following for permission to reproduce copyright material:

Figure 1.4 from Inside Europe in The *Guardian*, Guardian Newspapers Limited, (Black, I. 2003) © Guardian; Table 2.1 from *Rethinking Geopolitics*, Routledge (Taylor & Francis) (Ó Tuathail, G. and Dalby, S. (eds.) page 28, 1998); Figure 3.1 adapted from *The Geopolitics of Power and Conflict*, John Wiley & Sons, Ltd, (Nijman, J. page 688, 1992) © John Wiley & Sons, Ltd. Reproduced with permission; Figure 4.3 from *Rethinking Geopolitics*, Routledge (Taylor & Francis), (Ó Tuathail, G. and Dalby, S. (eds.) page 5, 1998); Figure 4.7 from *Reader's Digest*, March 1993, reprinted by permission of *Reader's Digest*; Figure 4.9 (cartoon FO 1110/392) by kind permission of the National Archives Image Library from 'Animal Farm' an adaptation (1951) of the book written by George Orwell; Figure 4.10 'If . . .' cartoons, The *Guardian*, copyright © Steve Bell 2000/All Rights Reserved; Figure 6.2 adapted from *Political Geography*, 2nd Edition, John Wiley & Sons, Inc., (an adaptation of the 'Indian Ocean' map from Glassner, M. I. 1996). Copyright © 1996 John Wiley & Sons. Inc. This material is used by permission of John Wiley & Sons, Inc.; Figure 7.3 Amnesty International poster, Amnesty International Publications, www.amnesty.org.

Figure 2.1 photo of the United Nations Security Council (UN/DPI Photo) and Figure 2.3 photo of the United Nations Secretary-General, Kofi Annan (UN/DPI Photo by Sergey Bermeniev by kind permission of the United Nations; Figures 4.1, 4.2, 4.5, 4.6, 6.1, 7.1, 7.4, 7.5, 8.1, 8.2, 9.1, and 9.2 photos: PA Photos; Figure 4.4 photo © Bettmann/CORBIS; Figure 5.1 photo ref: 108346 © Rex Features; Figure 8.4 photo © Christopher J. Morris/CORBIS.

Oxford University Press for an extract from *Globalization: A Very Short Introduction* by M. Steger 2003; and The Labour Party for an extract of the Prime Minister Tony Blair's speech *Doctrine of the International Community* to the US Economic Club on 22nd April 1999.

In some instances we have been unable to trace the owners of copyright material, and we would appreciate any information that would enable us to do so.

Chapter 1

REPRESENTING GEOPOLITICS

Key issues

- What is geopolitics?
- How does geopolitics function?
- What role do geopolitical visions play?
- How have American commentators made sense of the post-Cold War era?

What is geopolitics? Since the end of the nineteenth century, many scholars would answer this question with reference to the claim that geopolitics is traditionally concerned with the study of the state, its borders and its relations with other states (Heffernan 1998: 61). Given the prevailing nature of the international system (a world composed of nearly 200 states), this would appear to be a fairly reasonable starting position. Nation-states are very important and many conflicts have indeed occurred over the demarcation of territorial boundaries, the ownership of territory and access to resources such as oil and water. The last hundred years of human history would provide ample evidence such as the Israel–Palestine dispute and the long-standing tension between India and Pakistan over the ownership of Kashmir.

How does geopolitics work? One way to answer this question is to focus attention on the representation of geographical space. Geopolitics provides a way of seeing the world in which a great deal of emphasis is placed on exploring and explaining the role of geographical factors (such as territorial location and/or access to resources) in shaping national and international politics. In the process, ideas about places and populations are mobilised to construct 'geopolitical visions' (see Dijink 1996). While these visions of place can vary in cultural and geographical sophistication, the labelling of geographical space inevitably carries with it distinct implications for international relations and/or representations of national identities. As such many geopolitical writers have been preoccupied with providing insights for their own national governments and have frequently used geopolitical analysis to help make sense of the world. One of the most well known examples is provided by the former American secretary of state Henry Kissinger, who used the term 'geopolitics' to convey his thoughts about American foreign policy priorities in the midst of the Cold War struggle with the Soviet Union (Hepple 1986). But as we will see especially in Chapter 9, a new group of American neo-conservative intellectuals such as David Frum and Richard Pearle have been extremely influential in shaping the geopolitical priorities of the George W. Bush administration (Frum and Perle 2003).

There is a century-long tradition of viewing geopolitics as concerned not only with the activities of states but also with the 'big picture' of global politics. This is a highly visual approach to international politics because it frequently draws on maps to illustrate its findings. Although not everyone agrees with this basic concept of geopolitics, four features are generally accepted as typifying this kind of traditional geopolitical approach:

1. A way of seeing the world, often embellished as 'objective' and/or 'neutral'. In other words, the geopolitical writer positions himself (and it is usually men who are drawn to geopolitical theorizing) as detached and thus unburdened by ideology and/or prejudice.
2. A propensity to divide the world into discrete spaces often informed by a judgement on hierarchy, which positions some places and peoples as superior to others. Simple classifications (such as Heartland, Wild and Tame Zones, Land and Sea Power, Core and Gap) help to make apparent sense of a messy and deeply interconnected world. Maps play a crucial role in transmitting ideas about geopolitical spaces.
3. A desire to offer policy advice to states and their governments. Most geopolitical writers seek to use their 'God's eye' view of the world in order to formulate suggestions such as the development of new foreign and security policies.
4. Most geopolitical authors, unsurprisingly given point 3, openly display national partisanship. Their analyses of global geopolitics are thus approached from a very nationalistic point of view and rarely embrace cross-cultural and/or cross-regional understandings or perspectives (see Hepple 1986).

Are we witnessing the return of traditional geopolitics and the sorts of features identified above? While the Soviet Union and the **Cold War** have been apparently confined to history, traditional geopolitics is, according to the British academic and commentator John Gray, back in fashion as countries such as the United States and its allies warn of grave new dangers (Gray 2002 and see Glossary). Following September 11th, the legitimate use of military force occupies centre stage on the world political agenda, and geography appears to matter a great deal in terms of the location of new terror-related dangers and threats in regions such as Central Asia and the Middle East. In order to justify and legitimate new security policies, leading intellectuals attached to the Bush administration such as David Frum have drawn attention to the changing geographies of contemporary world politics (Frum 2003). As we shall see in Chapter 9, Frum was critical in persuading President Bush to draw clear boundaries between states fighting terror and those supporting terror. This demonstrates feature 2 of traditional geopolitics – a propensity to divide the world into discrete spaces.

Due to these shifting circumstances, new opportunities have emerged to reappraise the contemporary geopolitical situation. Given the nature of geopolitical theorizing, it is incumbent on us to note that all forms of political writing and interpretation are invested with values, conceits and prejudices (see Chapter 2). The remaining sections of this chapter consider a selection of contemporary geopolitical perspectives about the state of the world and the role of the United States in particular. The intention is not only to demonstrate how different writers interpret and represent contemporary geopolitics but also to reveal the intellectual underpinnings of their worldviews. The post-Cold War era (i.e. after 1989–90 and the disintegration of Cold War geopolitics) has been deliberately chosen because many commentators in North America, Europe and elsewhere remain intensely interested in representing and understanding the uncertain geopolitical condition of the world.

The key outcome from this chapter is to convince readers that geopolitical theorizing is never divorced from power–knowledge relations, and as a consequence there is actually no neutral or value-free way of viewing the world (contrary to feature 1 of traditional geopolitics).

Traditional geopolitics: the United States and the representation of global political space post-September 11th

For proponents of traditional and, as we shall see, more critical forms of geopolitics, these are unquestionably exciting as well as dangerous times. In the aftermath of the attacks on the United States on 11 September 2001, policy analysts, academics and journalists have been preoccupied with making sense of the changing geographies of world politics. This is the stuff of geopolitics.

The war against terrorism has apparently replaced communism as the catalytic motivation and ideological frame for a resurgent United States. As such the geographical representation of world politics has changed remarkably. The American-led coalition war against Saddam Hussein's Iraq in March 2003 is indicative of this trend even if it was not directed against the Al-Qaeda network itself. In the face of new dangers, the United States is apparently willing and able to use its immense military capacity to achieve rapid political results, often in a pre-emptive manner. Shortlived but arguably damaging to the domestic citizenry, the assault on the Iraqi war machine had the apparent advantage of removing the violent regime of Saddam Hussein, which had enjoyed American financial and military backing in the last decade of the Cold War. At that time, self-styled 'secular Iraq' was represented as a welcome and reliable bulwark against an Iranian theocracy described by most Western governments as 'fundamentalist'. After the invasion of Kuwait in August 1990, Saddam Hussein was identified by successive US administrations as the biggest threat not only to the Middle East but also, it could be argued, to resource-hungry America (Fig. 1.1).

A decade later, plans were afoot to remove him and his regime from the leadership of Iraq following multiple attempts, sometimes involving the United Nations, to dismantle his WMD programme. This transition illustrates only too clearly the changing geopolitical representation of threat as Iraq was transformed from an ally to a clear and present danger. Such a shift is not unique, however. From the perspective of the United States and its allies, the transition of the Soviet Union from wartime ally (against Nazi Germany) to Cold War adversary between 1945 and 1947 was arguably even more momentous.

Figure 1.1   A mural of Saddam Hussein found in a small café in Ortum, North West Kenya. When I asked the owner why Saddam Hussein was described as 'the hero', he explained that he admired the former Iraqi leader for 'standing up' to 'imperial America'
Photo: Klaus Dodds

Why do changing perceptions of threat matter? After September 11th and the overthrow of the Taliban regime in Afghanistan, Iraq was not the only state causing continuing concern to the United States. In his remarkable 2002 State of the Union address, President George W. Bush declared that Iran and Iraq alongside North Korea were part of an **'axis of evil'**. This is an excellent example of what we might call **practical geopolitics** (see Glossary). Bush's speech helped to actively reimagine global political space. No previous American administrations had visualized Iran, Iraq and North Korea in this interconnected manner. Associated with alleged encouragement for the development of WMD, terrorism and anti-American activities, this geographical label arguably provided a justification for the 2003 invasion of Iraq. Attention in Washington has now turned to Iran and North Korea (Iraq presumably having been removed from this category), while Syria has been warned not to give any support to terror groups or Saddam Hussein's fleeing political allies (see Chapter 9). Notwithstanding the very real concerns over US security, former vice president Al Gore, amongst others, has accused President Bush of squandering international sympathy following the September 11th attacks on New York and Washington DC.

Geographical labels such as 'axis of evil' matter greatly when they inform the foreign and security policies of the world's most powerful country, the United States. Military power has been deployed to excise a variety of threats identified by the US administration. As President Bush claimed in February 2002, geographical distance would provide no barrier or source of salvation to America's adversaries:

> Our armed forces have delivered a message now clear to every enemy of the United
> States: even 7,000 miles away, across oceans and continents, on mountain tops and in
> caves, you will not escape the justice of this nation (cited in Sardar and Davies 2002: 106).

While this vision of a border-free world clearly touches upon wider debates about how globalization is changing political, economic and cultural relations, debate continues to unfold as to the wisdom and efficacy of designating parts of global political space an 'axis of evil' (Sardar and Davies 2002, Frum 2003, Simpson 2003 and Chapters 2 and 9 below).

What is interesting about these geographical labels is the way in which the spectre of imperialism and colonialism is raised. Within the United States and elsewhere, critics argue over whether the world's first post-colonial state is similar in spirit and purpose to nineteenth-century imperial Britain or sixteenth-century imperial Spain. Is the United States an imperial state? This link with imperialism is important because modern expressions of geopolitics arguably were informed by the imperial worldviews of colonizing states such as Britain. As John Agnew (2002) has contended, the modern geopolitical imagination (based on a God's eye view of the world, for instance) is informed by a process and vision associated with European colonialism dating back to the fifteenth century (Pratt 1992 and see Chapter 2). Over the centuries, this imagination has come to assume that the world can be divided into civilized and primitive spaces (Young 2003). Thus labelling of political spaces during and after the Cold War arguably drew inspiration from earlier imperial points of view (Agnew 2002, 2003).

Even if the role of Great Powers and their opponents are not new (in terms of substance and terminology), these interventions in the early part of the twenty-first century come at a time of great uncertainty. After all, a deadly terror attack was shown to be only a plane ride away. But we also need to guard against ignoring less spectacular geo-economic processes shaping the contemporary condition. In our attempt to understand the unfolding geopolitics of terror, we should not underestimate the importance of the geopolitics of North–South relations (see Chapter 3). Global capitalism may be the preferred method of economic organization but it is also, arguably, inherently geographically divisive. Instead of promoting free and fair trade in conjunction with peaceful international relations, critics contend that the most powerful states such as the US (and European Union states) frequently revert to protectionist trade politics in order to promote their own self-interests. In other words, powerful states promote their own unilateral geopolitical and geo-economic priorities.

Whether or not one shares the same viewpoint as President George W. Bush and his neo-conservative intellectuals, the new century has already provoked some substantial debates about the future role of the United States and the shape of global geopolitics. Four recent viewpoints (well known in the United States, Europe and elsewhere) will facilitate our investigation of traditional and more critical forms of geopolitics and will serve our purpose of exploring a little further how global political space is being geographically constituted and contested.

Viewpoint 1 – the end of history v. the clash of civilizations

A number of commentators in the United States have attempted to explain the economic, geopolitical and cultural significance of the ending of the Cold War. The discussions go to the heart of geopolitics because they are concerned with the political understanding of the world. The purpose of these reviews is to demonstrate that geopolitics should be understood as a project dedicated to the production, circulation and interpretation of global political space. Different forms of geopolitical interpretation are illustrated using the well-known commentaries of Francis Fukuyama and Samuel Huntington concerning the passing of one geopolitical order and the emergence of a new (if uncertain) geopolitical order. While Fukuyama's thesis is decidedly more optimistic on the fate of the United States and the West in the new world order, Huntington's arguments on the interactions between civilizations are laced with apprehension and scepticism about the deterritorializing world order (see O Tuathail and Luke 1994, O Tuathail 1999, Luke 2003). However, the major theme unifying these two rather different commentaries is their profoundly *anti-geographical* aspect, which tends to ignore the complex geographies of world politics. This is an important point, because despite protestations to the contrary, modern geopolitical viewpoints in their desire to divide the world into large regions or zones, often underestimate complexity and interconnection. In one sense they have to eschew such complications, otherwise their maps and/or visions of global political space would be compromised. For example, as with other influential American commentators such as Robert Kaplan (2000), diverse cultural regions of the world are frequently labelled as either barbaric or threatening to the interests of a great power such as the US, thereby underestimating (perhaps deliberately) the inherent problems and issues in visualizing a rapidly changing world.

The end of history and the triumph of the West?

The most famous contribution – in part because of the timing of its publication and the title – was the unashamedly triumphalist essay by Francis Fukuyama (1989) entitled 'The end of history' hailing the collapse of the **Berlin Wall** as a victory for the United

States and a defeat for the forces of communism and tyranny. According to Fukuyama, the post-Cold War era would witness a transformation of regions such as Eastern Europe from state-managed communism towards liberal democracy and market economics. In his book, *The End of History and the Last Man*, Fukuyama returned to this theme of the triumph of liberal democratic political and economic systems:

> The most remarkable development of the last quarter of the twentieth century has been the revelation of enormous weaknesses at the core of the world's seemingly strong dictatorships, whether they be of the military-authoritarian Right, or the communist-totalitarian Left. From Latin America to Eastern Europe, from the Soviet Union to the Middle East and Asia, strong governments have been failing over the last two decades. And while they have not given way in all cases to stable liberal democracies, liberal democracy remains the only coherent political aspiration that spans different regions and cultures around the globe (Fukuyama 1992: xiii).

Moreover, Fukuyama's thesis was predicated on the assumption that liberal democracy in alliance with market economies had the capacity to fulfill basic human needs of self-worth and material well being. The consequences for the post-Cold War world would be:

> The creation of a universal consumer culture based on liberal economic principles, for the Third World as well as the First and Second. The enormously productive and dynamic world being created by advanced technology and the rational organization of labor has a tremendous homogenizing power . . . The attractive power of this world creates a very strong disposition for all human societies to participate in it, while success in this participation requires the adoption of the principles of economic liberalism (Fukuyama 1992: 108).

However, Fukuyama's thesis on the global **hegemony** of liberal democracy and market economics has been heavily criticized by academic and political commentators who consider his arguments regarding the so-called Second and Third Worlds as geographically and historically insensitive. While democratic governments have emerged in parts of Southern Africa, South America and South East Asia, their long-term viability remains open to question given the repressive nature of administrations in Indonesia, Malaysia, Peru, Zimbabwe and South Korea. It is not inevitable that fledgling democratic states will follow the pre-determined pathway (as modernist development theories assumed in the past); previously authoritarian regimes may succumb to the pressures of economic liberalism and the demand for greater political freedoms, democratic elections, formal political participation and regular elections. With a population of over one billion people, China will also need to be scrutinized closely to see whether greater political freedom accompanies moves towards market liberalization.

Fukuyama's arguments frequently underestimated the differences between liberal and democratic governments. The current administrations of countries such as Britain, Chile, Japan, the USA and India are characterized by different forms of institutions, participation and practices. The assertion that the West has been triumphant ignores the interplay of different cultures and civilizations, which might lead to greater exchange and hybridity rather than unproblematic global subservience to ideas of democracy and market economies.

Global cultures and the clash of civilizations?

By contrast Huntington's (1993) paper 'The clash of civilizations', published in the American journal *Foreign Affairs*, was pessimistic rather than optimistic about the ending of the Cold War. Huntington argued that the new global order would be characterized by the interaction of seven or eight large civilizations – Sinic (Chinese), Hindu, Islamic, Japanese, Latin American, Orthodox, Western and possibly African (see Fig. 1.2). As the West is not considered to be culturally and politically dominant, the **new world order** will witness the growing influence of Islamic, East Asian and Chinese civilizations. The expanding Chinese economy has already contributed to the growing political confidence of the People's Republic of China. The handover of Hong Kong in July 1997 and the militant attitude (condemned by the Western powers) displayed towards the pro-independence forces in Taiwan from the mid-1990s onwards are recent illustrations of China's political and cultural self-confidence. According to Huntington, the implication of such a development is that the West's capacity to manage world politics will be challenged by the Sinic (Chinese) and Japanese civilizations in the next century. In support of this proposition, Huntington presented evidence of a growing trend of anti-Western sentiment in Islamic and Asian countries in spite of the fact that the Western-educated elites of these countries often speak English and are anxious to conduct business with international organizations.

Huntington's rejection of the Fukuyama thesis was reiterated in his book-length study entitled *The Clash of Civilizations and the Remaking of the World Order* (1996). As in his earlier paper, he argues that the major divisions in the new global order will be based on the exchanges between civilizations rather than ideologies. For Huntington, the Cold War was a brief moment of international order in a longer historical context of confrontations and tensions. In the aftermath of the Cold War, international order will, therefore, be characterized by the return of civilizational tension as core states such as China, Japan and the United States seek to either expand or preserve their influence around the globe. His message for Western audiences is that the West will have to embark on a strategy which not only strengthens political and cultural values between Western civilizations but also seeks to construct new alliances with other civilizations (see Fig. 1.2).

The dangers of geographical generalization

As with Fukuyama's 'end of history' thesis, Huntington's arguments rely upon a series of sweeping generalizations about the state of world politics and processes such as **globalization**. The apparent threats posed by non-Western civilizations are frequently exaggerated and ignore the complexities posed by the apparent rise of fundamentalist movements in major religions such as Islam. Huntington's use of the

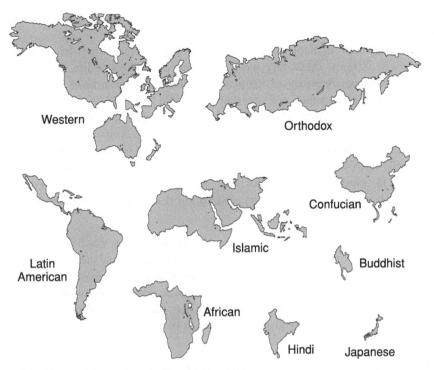

Figure 1.2 The world according to Samuel Huntington

recent disintegration of Russia and Yugoslavia to identify a tendency for multicivilizational societies to fracture along civilizational faultlines is, to this author, a deeply flawed
argument. While the example of post-apartheid South Africa provides evidence that
multicivilizational societies are not inevitably doomed, the destruction of Yugoslavia
between 1992 and 1995 cannot simply be reduced to the interplay of different civilizational forces (Islam, Orthodox Christianity and Western Christianity) because that
would at the very least ignore what some have seen as Western complicity in this destruction (O Tuathail 1996, Ali 2000). In a related vein, Huntington's characterization of
the Latin American civilization as a backward if reliable appendage of the Western
civilization effaces some of the important changes occurring within and between the
Western and Latin American civilizations: large migratory movements of people and
capital, hemispheric integration within the Americas and the growing inequalities
between migrants and the settled population. The notion of a 'clash of civilizations'
simply fails to capture the nature and intensity of interactions and exchanges between
cultural regions. Huntington's all-encompassing definition of a civilization as 'the
highest cultural grouping of people and the broadest level of cultural identity people
have short of that which distinguishes humans from other species' has the effect of reducing people and events down to broad and essential differences (cited in O Tuathail

1996: 244). In other words, generalizing about political space and cultures is a difficult proposition.

In the aftermath of September 11th, it is perhaps unsurprising that Huntington's 'clash of civilizations' thesis received a new lease of intellectual life. Given that self-confessed Islamic extremists used aeroplanes to commit an extraordinary series of assaults on the United States, proof was apparently available that the West and Islam were on a collision course. In responding to such a suggestion, the late Palestinian-American academic and activist Edward Said (2001) warned in the *Nation* magazine of a 'clash of ignorance'. He notes that Huntington's description of the Islamic civilization relies on a heavily simplified notion of identity and culture. The plurality of the Islamic world is negated in favour of a one-dimensional view of regions, cultures and peoples. Said also takes Huntington to task for the assumption that he enjoys an Olympian view of the world:

> More troubling is Huntington's assumption that his perspective, which is to survey
> the entire world from a perch outside all ordinary attachments and hidden loyalties, is
> the correct one, as if everyone else were scurrying around looking for answers that he
> has already found. In fact, Huntington is an ideologist, someone who wants to make
> 'civilisations' and 'identities' into what they are not – shut down, sealed off entities
> that have been purged of the myriad currents and counter currents that animate human
> history . . . (Said 2001: 10).

Fukuyama and Huntington's worldviews are geopolitical in the sense that they share a common concern for the mapping of global geopolitical space. Fukuyama, as a former member of the Department of State in Washington, advised on American foreign policy under the Reagan and Bush Senior administrations, while Huntington's lengthy career spanned elite universities and political foundations such as the Council for Foreign Relations. Their commentaries formed a geopolitical discourse, which constructed international affairs as a dramatic stage for competing ideas and states. But they also, as Said warns with reference to Huntington, exhibit an intellectual conceit which allows them to believe that it is possible to simply divide the world into discrete spaces without one's personal (often unstated) prejudices and assumptions influencing this mapping.

Viewpoint 2 – pax Americana: the United States and empire

For a country founded on a revolution against eighteenth-century imperial Britain, it is perhaps surprising to see the United States accused by some observers of being a latter-day imperial power (Agnew 1983, Countryman 1985). As a co-founder of the United Nations, the United States played a pivotal role in ensuring that all forms of territorially based imperialism were brought to an end in favour of the principle of self-determination being extended for all peoples. In contrast to imperial Britain and France, its acquisition of overseas territories was relatively modest (the Philippines, Cuba and island chains throughout the Pacific Ocean) in favour of continental expansion in the nineteenth century. As the United States stretched its frontier from the original states on the Eastern seaboard to new lands across the Mid West and the Rockies, indigenous Indian populations, Mexico and Southern states such as Georgia succumbed to disease, war and the general process of nation building. By the time the modern United States was created in the late nineteenth century, the mythology surrounding its creation was well and truly entrenched, even if some famous authors such as Mark Twain were denouncing 'American imperialism' (Campbell 1992).

A mixture of theology and frontier nationalism, imbued with a strong sense of moral righteousness, produced a heady potion in this new nation (Agnew 1983, O Tuathail and Agnew 1992). The United States, as the world's first post-colonial state, was seen as an example to the rest of the world. This sense of political confidence was underwritten, as any US dollar bill will confirm, by a belief that 'In God we trust'. Moreover, this perception of distinctiveness was also territorialized in the sense that the 'new world' was compared favourably to the decadent 'old world' of imperial Europe. America was the land of opportunity and freedom while Europeans remained imprisoned by class, wealth and privilege (see Viewpoint 3, below).

During the Cold War, the Soviet Union stood accused of being, in the words of President Ronald Reagan, the 'evil empire' in contrast to the Western European nations such as Britain and/or France. The incantation of a struggle between good and evil is a well-entrenched feature of presidential rhetoric and suggests that places can be linked to expressions of national identity. As President Eisenhower noted in his inaugural address in January 1953:

> We sense with all our faculties that forces of good and evil are massed and armed and opposed as rarely before in history. This fact defines the meaning of this day (cited in Appy 2000: 3).

Thus, for many Americans, it would be something of a shock for the United States to be charged with being 'imperial' (given its anti-colonial inception), let alone being accused of insularity, inequality and/or insolence (Cox 2002).

The Canadian broadcaster and academic Michael Ignatieff, in his new book *Empire Lite*, provides a sympathetic examination of the dilemmas facing the world's remaining superpower (Ignatieff 2003). Part of the problem, Ignatieff contends, lies with the longstanding American reluctance to see itself as an imperial power with corresponding global responsibilities:

> The Roman parallels are evident, with the difference that the Romans were untroubled by an imperial destiny, while the Americans have had an empire since Teddy Roosevelt, [i.e. turn of the twentieth century and coincidental with the US occupation of the Philippines] yet persist in believing they do not (Ignatieff 2003: 1).

Writing in the aftermath of September 11th and the US-led invasion of Afghanistan in 2001, Ignatieff suggests that the United States will need to foster a new self-image: a benevolent imperial power with a global sphere of influence but without formal colonies. As the world's largest military power, the United States remains the only country capable of launching military and humanitarian action in diverse locations such as Afghanistan, Kosovo and/or Bosnia. While European countries reduced their defence spending after the Cold War, the United States increased its defence budget to levels not previously witnessed since the most intense phases of the Cold War.

In this 'new world of geopolitics' (Ignatieff 2003: 20), the United States has no choice but to play a more active role in shaping global political developments (informed by a new geopolitical vision). Isolation is not considered to be an option. Given Al-Qaeda's stated aim of removing the United States military presence from the Middle East and Islamic World and Western cultural influence more generally, the US has no choice but to intervene in order to confront a never-ending form of global threat. The global eradication of terror in combination with humanitarian/political crises in seemingly faraway places such as Kosovo (1999) and Afghanistan (2001) provides a necessary and just pretext for 'imperial' America. In these uncertain times, a strong United States is seen as an imperative and the US may have to indulge in 'temporary imperialism' in order to help construct post-civil war democracies in Afghanistan and now Iraq. Ignatieff's conclusion runs counter to the Bush administration's belief that national security will only be assured if the US is prepared to mount pre-emptive assaults anywhere in the world. However, US imperialism (of a more temporary nature) could not only be a force for international order but also remain necessary to the maintenance of a global human rights culture.

Sardar and Davies (2002) in their book *Why do People Hate America?* present a rather different view of 'imperial' America. The United States is not only widely resented for its military, political, economic and cultural dominance but also stands accused of perpetuating double standards. Thus, for example, the United States frequently champions democracy and the pursuit of freedom while at the same time actively disrupting the democratic activities of other states (see chapters 3, 7 and 9). During the Cold War, the US intervened in the implementation of democratic elections in Italy (1948–1970s), Lebanon (1950s), British Guiana (1950s–1960s), Indonesia (1965), Chile (1964–1970), Nicaragua (1984), Panama (1984 and 1989), Haiti (1987–1988) and Kosovo (1998). Elsewhere, the continued uncertainty over a Palestinian nation-state is frequently cited as an illustration of how the United States is prepared to generously support the Jewish state of Israel but not an Arab-administered Palestine. What worries many onlookers is that America will use the contemporary 'war on terror' as a pretext to either intervene or support the right of others to carry out anti-terror operations (for example, the Russian anti-terror campaigns in Chechnya) around the world. 'Imperial' America (and accompanying geopolitical representations) is thus legitimated by the never-ending struggle against terror movements and alleged sponsor states such as Iraq, Syria and Libya.

Many fear that the declaration of a 'war on terror' is being used in an opportunistic manner. Since its creation in the eighteenth century, the self-identity of the United States has been strongly linked to war. The struggles over the American frontier became an all-important element in the founding mythology of America. The subsequent struggle between American capitalism and Soviet-led communism created a new global frontier during the Cold War (Campbell 1992).

The 'war on terror' rests, it is opined, on two older myths of American political and cultural development. The first involves the myth of the 'savage war' in which the United States and a white Christian civilization more generally are imperilled by primitive tribes. The second involves the invocation of the 1941 Japanese attack on the US naval base at **Pearl Harbor** and the belief that America is implicated in a 'good war' because it was attacked in an unprovoked and unjustified manner (Sardar and Davies 2002: 190–1 and see Glossary). Both myths can be and were used in the aftermath of September 11th to justify belligerent American activities in Afghanistan and Iraq. What these debates about 'imperial America' suggest is that representations of global political space do not exist in a geographical and historical vacuum. They help shape expressions of national identity and national purpose.

Viewpoint 3 – power and persuasion: Europe v. the United States

During the Cold War, Western Europe in the form of the European Economic Community (now the European Union) and the North Atlantic Treaty Organization (NATO) lay at the heart of the 'Western Alliance' led by the United States (Fig. 1.3).

With the possible threat of a conventional and/or nuclear attack by the Soviet Union and its Warsaw Pact (the equivalent of NATO, created in 1955) allies, geopolitical and cultural solidarity between Europe and the United States was considered to be of the greatest importance. This sometimes demanded that Western allies agreed in the main not to publicly pursue disagreements over military strategy or political interventions for fear of providing the Soviet Union with evidence that the 'West' was weak and divided. The decision by the French to leave the US-led command structure of NATO in 1966 was therefore considered shocking by the Americans and the British. Likewise the Soviet Union was ruthless in its elimination of any signs of divisions among the Warsaw Pact members. The prevailing geopolitics of the Cold War provided a powerful incentive on both sides to avoid schisms.

More recently, however, the United States administration of President Bush and European Union/NATO partners have publicly disagreed over the 2003 war against Iraq. Furious at the public opposition of France and Germany to plans for invasion, US Defence Secretary Donald Rumsfeld in February 2003 dismissed these countries as part of an 'Old Europe'. By way of contrast, he pointed to the generous political support given to the United States by ex-Warsaw Pact countries and/or new member states of NATO in the former Eastern Europe bloc such as Poland, Hungary, Bulgaria and Estonia (Fig. 1.4).

These, according to his pointed geopolitical analysis, were part of a 'New Europe'. France, Germany and Russia (even if Russia remained an ally with the US in the 'war on terror') could not be relied on, and as part of a broader military readjustment the United States began to relocate its forces from Germany and Belgium to Poland and Bulgaria. Britain, because of its long-standing support of the United States, was aligned with a 'New Europe'.

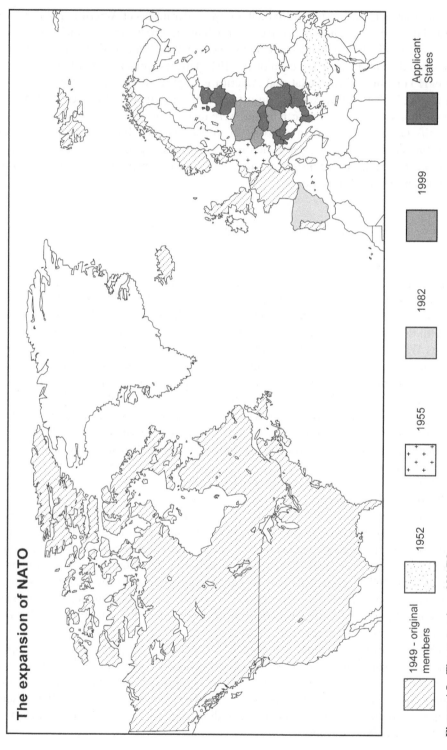

Figure 1.3   The expansion of NATO

Figure 1.4 Post-Iraq: how the splits emerged
Source: The *Guardian*, 28 April 2003 © Guardian

According to American political commentator Robert Kagan, however, America's spat with European Union partners such as France is more than simply a 'falling out' over the subsequent treatment of Saddam Hussein's Iraq. In his widely publicized article 'Power and weakness' and subsequent 2003 book *Paradise and Power*, he contends that competing views of power and representation lie at the heart of the rupture. The 'European' response to international crises such as Iraq and the where-abouts of WMD was to instinctively seek a multilateral solution involving the United Nations. Underwriting this political view is a belief that the world should be shaped by international co-operation and negotiation rather than by military force. This view of power is shaped by the intellectual vision of Immanuel Kant and the 'perpetual peace'. The United States, meanwhile, naturally turns to unilateral solutions because it con-siders international organizations and international law to be a chimera. This view of power is informed by the writings of Thomas Hobbes and his vision of a violent 'state of nature' (see In focus 1.1).

In focus 1.1: Immanuel Kant and Thomas Hobbes

Immanuel Kant was an eighteenth-century philosopher who wrote a pamphlet entitled *Perpetual Peace*. In this work Kant contends that in a world composed of constitutional and/or republican states, it was possible to conceive of a state of 'perpetual peace'. With a shared political and legal culture, international co-operation could thrive rather than there necessarily being a descent into anarchy and war. Kant's vision of a world based on peaceful co-existence inspired later theorizing on the role of democracy in promoting peaceful international relations as opposed to simple republicanism. It has been contended that President Woodrow Wilson was deeply influenced by this body of thought when he promoted a vision of a world regulated by a League of Nations in the aftermath of the Second World War.

Thomas Hobbes was a seventeenth-century political and legal theorist who contended that social and political life, under the 'state of nature', is 'solitary, poor, nasty, brutish and short'. In order to promote the good life, it was essential to foster the development of a strong government cum civilization, which could protect a national territory; a domestic population allied with secure property relations. The sovereign state was thus seen as something of an escape route from the unpleasant 'state of nature'. Hobbes's ideas have contributed to a body of political thought called *Realism*, which is considered further in Chapter 2.

Writing in June 2002, Kagan propounds:

> It is time to stop pretending that Europeans and Americans share a common view of the world, or even that they occupy the same world. On the all-important question of power – the efficacy of power, the morality of power, the desirability of power – American and European perspectives are diverging. Europe is turning away from power, or to put it a little differently, it is moving beyond power to a self-contained world of laws and peace and relative prosperity. . . . The United States meanwhile remains mired in history, exercising power in the anarchic Hobbesian world where international laws and rules are unreliable and where true security and the defence and promotion of a liberal order still depend on the possession and use of military might (Kagan 2002: 1).

This is a controversial argument and at its heart it relies not only on a fairly crude assessment of 'power' but also on two major contentions. Power in this context is viewed in terms of force and military capability rather than as part of a knowledge–power nexus that would be familiar to followers of Michel Foucault (see Chapter 2). Kagan contends that the strategic perspective of Europeans has been shaped by a widespread desire to promote peace and stability within and beyond Europe. Europeans are, in his words, 'the preachers' as opposed to the US 'sheriff'. The heavy human cost of the Second World War in combination with the regional development of the European Union has given credence to the view, which is instinctively sympathetic to diplomacy and conflict resolution rather than armed intervention. Despite the loss of many American lives during the Second World War, the United States and its mainland civilian population was never directly bombed. It also never experienced the trauma of foreign occupation during and after 1945. Austria and Germany, by contrast, were occupied by the Allied forces of the Soviet Union, the United States, Britain and France until 1955.

More controversially, Europeans are also in favour of multilateral solutions to conflicts because they do not possess the military capability to impose their will. In other words, Europe's military weakness means that Europeans are more likely to tolerate 'threats' and 'dangers'. Unsurprisingly, perhaps, some American commentators described French and German critics of the US-led invasion of Iraq in March 2003 as part of the 'axis of weasels' and rebranded 'French fries' as 'Freedom fries'.

The earlier (lack of) response to the disintegration of Yugoslavia in the 1990s highlighted still further the different approaches to power politics amongst Europeans and Americans. Despite its geographical proximity to the conflict, the European Union (EU) stood accused by the United States of failing to end hostilities, which cost over 250,000 lives and led thousands more to flee for their lives. For many Americans and also European observers, the EU was exposed as an impotent body incapable of organising sufficient military forces and/or humanitarian assistance (Simms 2001). In July 1995, in one of the bleakest periods of this confrontation between rival factions, the town of Srebrenica was overwhelmed by Bosnian Serbian forces and some 7,000 Bosnian Muslim men and children were murdered, while Dutch peacekeepers in close proximity were powerless to intervene. As one political geographer later noted:

> Yet geography made the violence of Srebrenica unique in two ways. The first was its location in Europe. . . . What was happening in Srebrenica was close both geographically and visually to 'us', to the safe and civilised world of the European Union. . . . Srebrenica was also special because it had been declared by the United Nations a 'safe area' in April 1993. Designating the town as a 'safe area' represented an effort by the international community to legislate a special zone of order and security amidst the generalised disorder and warfare in Bosnia (O Tuathail 1999: 120–1).

The moral bankruptcy of the United Nations and EU states was deeply exposed and it was not until NATO forces led by the United States intervened that a peace settlement was eventually secured and signed in Dayton, Ohio in December 1995. The location of the peace settlement was highly significant as the United States demonstrated that the European Union was incapable militarily and/or politically of ending this European crisis. Subsequently, American administrations (both Democrat and Republican) have expressed frustration with Europe's reluctance to confront international problems, including the subsequent violence in the former Yugoslav region of Kosovo, in a direct manner. The strategy of the Europeans, according to Robert Kagan, is to try and constrain the United States through international negotiation while at the same time refusing to share the burden of protecting international political order.

From the United States' strategic perspective, however, Europe seeks to constrain precisely because it lacks the military and political determination to respond to threats and dangers. Moreover, unlike Europe, the United States as the world's greatest superpower faces a far greater series of threats. Saddam Hussein's Iraq was just one of many polities that sought to undermine US military, political and cultural power. It is no accident that the Al-Qaeda network targeted the American cities of New York and Washington DC rather than the European cities of Paris, Rome or Berlin.

Given these different understandings of power, Kagan suggests that European countries are unlikely to radically review their defence spending and in practice their Kantian view of power. The development of a European armed force will take time, not least because EU membership was substantially increased in May 2004. It is difficult to imagine the EU exercising its international political muscles in the midst of a period of geographical expansion as former Eastern European states and others are embraced. In Kagan's view, the United States as the world's largest superpower will have to change its view of the Europeans and their concern about unilateral excesses:

> Americans are powerful enough that they need not fear Europeans, even when bearing gifts. Rather than viewing the United States as a Gulliver tied down by Lilliputian threads, American leaders should realise that they are hardly constrained at all, that Europe is not really capable of constraining the United States. If the United States could move past the anxiety engendered by this sense of constraint, it could begin to show more understanding for the sensibilities of others, a little generosity of spirit. It could pay its respects to multilateralism and the rule of law and try to build some international political capital for those moments when multilateralism is impossible and unilateral action unavoidable (Kagan 2002: 17).

This appeal for a 'little common understanding' appeared in the summer of 2002 and thus before the build-up to the 2003 US-led assault on Iraq. The arguments over the management of Saddam Hussein's Iraq demonstrated only too clearly that the Bush administration viewed the dissenting voices in the United Nations as unhelpful obstacles.

Moreover, given the American treatment of the 'unlawful combatants' captured in the 2001 attack on Afghanistan, it is perhaps unsurprising that many European commentators and governments including France, Belgium and Germany were critical of the unilateral actions of the United States. International law was apparently put to one side as the George W. Bush administration invented the category 'unlawful combatant' and shipped prisoners thousands of miles to Cuba, where they have been denied legal representation. Coupled with its refusal to ratify provisions for the International Criminal Court in Rome, this contributed to a widely held European belief that the United States was uninterested in securing a more rules-based international political and legal order. The US condemned Iraq for its unilateral behaviour over WMD development while at the same flouting international conventions (such as the 1925 and 1949 Geneva Conventions) designed to restrain the excesses of all states.

Viewpoint 4 – globalization and the 'ozone hole': the Core and the Non-Integrating Gap

In an article for the popular magazine *Esquire*, Thomas Barnett of the US Naval War College outlined his controversial argument concerning the strategic wisdom of the 2003 assault on Iraq. As a sometime advisor to the Department of Defense (the Pentagon),

Barnett argued that the United States needs a new operating theory of the world in the wake of the Cold War and the September 11th attacks. At the heart of this new theory should be a concern with 'disconnectedness' because the 'rogue regime' of Saddam Hussein was insufficiently 'connected' with globalization and its rules, norms and the ties that bind countries and cultures together. Barnett contends that the world can thus be divided into two major constituencies:

> Show me where globalisation is thick with network connectivity, financial transactions, liberal media flows and collective security, and I will show you regions featuring stable governments, rising standards of living, and more deaths by suicide than murder. These parts of the world I call the Functioning Core, or Core. But show me where globalisation is thinning or just plain absent, and I will show you regions plagued by politically repressive regimes, widespread poverty and disease, routine mass murder, and – most important – the chronic conflicts that incubate the next generation of global terrorists. These parts of the world I call the Non-Integrating Gap, or Gap (Barnett 2003: 2).

Armed with this simple binary division of the world, Barnett argues that the next generation of US military missions will be exclusively located in the 'Gap' or, if you prefer, globalization's 'ozone hole'. The entire 'Gap' is thus represented as a substantial 'strategic threat environment' precisely because it shows little willingness (or perhaps capacity) to adopt the global rule-set of democracy, transparency and free trade (Fig. 1.5).

Geographical membership of the Core and the Gap is the most intriguing part of Barnett's thesis. The division of global political space is further refined. Although he is cautious to avoid the claim that certain cultures and societies are somehow incapable of embracing democracy and the global economy, he defines the Core states as North America, Latin America in the main, the European Union, Russia, Japan, Australia, New Zealand, South Africa and Asia's emerging economies including India, China and South East Asian states. The Core is not simply composed of what used to be known in development circles as the 'North'. Russia under President Putin and the reforming governments of China are viewed as demonstrating a willingness to embrace the values implicit in globalization.

The 'Non-Integrating Gap' is defined as those regions currently excluded from the 'Core' of globalization, namely the Caribbean, Africa, the Balkans, Central Asia and the Caucasus, the Middle East and South West Asia. Two billion people are thus defined as outside the core values and practices of globalization. Fifty years earlier some of these countries would have been described as 'underdeveloped'; now they are represented as part of a 'Non-Integrating Gap'. Given the high levels of interconnectivity that exist between regions and cultures (whether the US and the 'Core' like it or not), Barnett argues that ignoring the backwaters of globalization is not an option:

> If we draw a line around the majority of those military interventions, we have basically mapped the Non-Integrating Gap. Obviously, there are outliers excluded by this simple approach, such as an Israel isolated in the Gap, a North Korea adrift within the Core, or

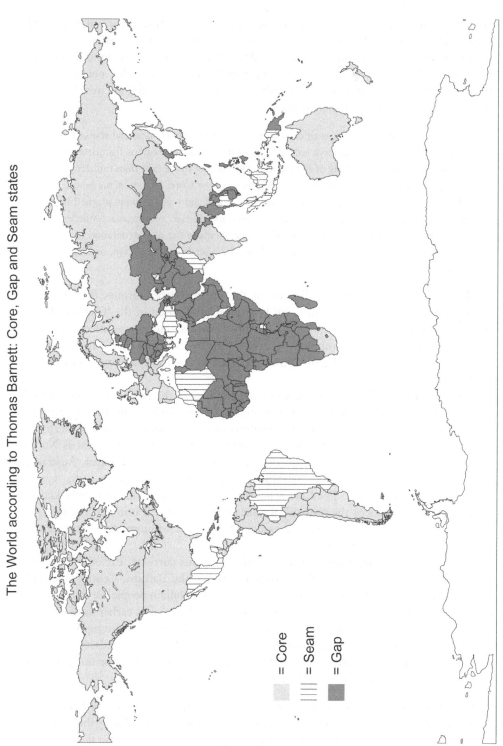

Figure 1.5 The world according to Thomas Barnett

a Philippines straddling the line. But look at the data, it is hard to deny the essential logic of the picture: If a country is either losing out to globalisation or rejecting much of its content associated with its advance, there is a far greater chance that the US will end up sending forces at some point. Conversely, if a country is functioning within globalisation, we tend not to send our forces there to restore order to eradicate threats (Barnett 2003: 3).

Alongside other members of the US military community, Barnett believes that these new geopolitical schisms have been brought into sharp relief by the 11 September 2001 attacks. While the practice of asymmetrical warfare is not new, the willingness of a world-wide terror network to inflict suicidal warfare on US military and economic interests is arguably unprecedented. Americans are now forced to endure new restrictions on their freedom and the apparatus of government was reformed to encompass the establishment of a new Department of Homeland Security.

The key dimension to this pessimistic assessment is the belief that the Al-Qaeda network was sustained in the 'Gap', where weak states and a general condition of lawlessness provide ideal breeding grounds for terrorism. Hence, the weak and disconnected states of Sudan and Afghanistan provided an ideal location for Al-Qaeda's training bases. Responding to the threat posed by Al-Qaeda requires the United States to tackle the flows and networks that enable people, weapons and terror operations to move across the boundary between the Core and the Non-Integrated Gap. The geopolitical frontline is defined by so-called 'Seam States' that provide access to the Core precisely because they geographically and politically straddle the functioning and non-functioning areas of the global economy and polity. Barnett identifies a number of 'Seam States': Mexico, Brazil, South Africa, Morocco, Algeria, Greece, Turkey, Pakistan, Thailand, the Philippines and Indonesia. Little evidence is provided to justify this eclectic selection. Some of the states have been defined as predominantly Muslim populations (Indonesia, Turkey and Pakistan, Morocco and Algeria) while others might be identified as overwhelmingly Christian (Mexico). Some have endured attacks carried out by Al-Qaeda (Morocco and Pakistan) while others have been embroiled in costly civil wars involving Islamic terror groups (Algeria and the Philippines). States such as Mexico and Greece appear to be listed just because they happen to be geographically proximate to the United States and Western Europe respectively. Interestingly, most of the so-called 'Seam States' could reasonably be described as functioning if vulnerable democracies.

The implication of this new global geopolitical map of Core, Non-Integrating Gap and Seam States is immense. Barnett contends that the US must adopt a more interventionist set of policies in regions such as the Middle East and Sub-Saharan Africa. Isolationism is not a geopolitical option. America must be prepared to intervene in order to reform authoritarian governments which help or merely fail to hinder the export of terror. Establishing new military bases in combination with active political intervention is held to provide a new model for responding to the threat posed by Al-Qaeda and its supporters. The United States, Barnett insists, should use its massive military capability to confront suicidal terrorism head-on in the Gap rather than waiting for

Al-Qaeda to launch another attack from its bases or areas of influence in the Non-Integrating Gap or Seam States. The eventual eradication of terror would then allow for the envisaged exportation of a more intense form of globalization. In Barnett's words:

> I know most Americans do not want to hear this, but the real battlegrounds in the global war on terrorism are still *over there*. If gated communities and rent-a-cops were enough, September 11 never would have happened. . . . We ignore the Gap's existence at our peril, because it will not go away until we as a nation respond to the challenge of making globalisation truly global (Barnett 2003: 5 with original emphasis).

In a manner remarkably similar to politically conservative interpretations of the Cold War, America is conceptualized as simply responding to a new global threat (communism in the past, terror in the present and future). The United States is thus apparently disconnected from the active production of global divisions and inequalities that perhaps help to generate phenomena such as the Al-Qaeda network.

As we shall see in Chapter 9, Al-Qaeda was a product of the Cold War struggle between the United States and the Soviet Union as well as important regional conflicts such as the Arab–Israeli dispute. When the Soviets invaded Afghanistan in 1979, the Americans viewed this intervention with considerable alarm given the country's proximity to the Middle East. The Reagan administration (and the Saudi Arabian government) therefore supplied monies and weapons to the anti-Soviet resistance movement, the Mujaheddin. With the Soviets successfully removed from Afghanistan in February 1989, the well-armed Mujaheddin then contributed not only to the creation of the Taliban regime in Afghanistan but also to the Al-Qaeda network. Both were intent on removing all traces of Western influence (including military bases) from the Middle East and the Islamic world. Thus, Barnett's proposed strategy, if fully implemented (the number of American military bases located around the world is rising), would arguably worsen anti-American sentiment, especially if the Palestine question remained unresolved.

While it may be convenient to claim that the Non-Integrating Gap is somehow profoundly disconnected from mainstream globalization, many of these countries and regions with their long experience of European and even American colonialism have experienced (often in a manner not of their collective choosing) cultural, economic and political interconnection and interpenetration. If there is ambivalence about globalization (and many Americans are also ambivalent, see Chapter 8) it is rooted in a concern that the United States and its allies (whether fellow governments or multinational companies) are able to exercise a disproportionate influence in the so-called Gap. Even if we accept Barnett's terminology, the Core and the Non-Integrating Gap are deeply interrelated and thus globalization is more complex than a mere adoption by the rest of the world of the preferences of the United States.

Barnett's analysis, like that of the other writers mentioned in this chapter, is indicative of a traditional geopolitics in so far as it proffers policy advice to his national government.

Conclusions

The United States, as the world's largest military superpower, does face some extra-ordinary representational challenges. How does it reinterpret global political space in the absence of the Cold War adversary the Soviet Union? How does it respond to the ongoing yet 'fluid threat' posed by the Al-Qaeda terror network? And when is intervention simply either an act of unwelcome imperialism or illustrative of paro-chial self-interest? Geopolitical writers, amongst others, have been anxious to provide answers to these kinds of question and as such have also sought to construct new imaginative geographies of global politics with associated implications for military intervention.

Geographical representations of this kind do not occur in a political vacuum. The United States was condemned alongside the United Nations for not intervening in the case of the Rwandan genocide in 1994 (because it considered Central Africa of low strategic salience) and was condemned for an imperial-style intervention in Iraq in 2003 (because it considered Iraq of high strategic value). Without US intervention in Palestine–Israel (so many believe on both the political Left and the Right), it is unlikely that this particular long-standing territorial dispute will be resolved given US financial and military support to Israel. However, the wider American public (only 14 per cent of whom possess a passport) may appear introverted and poorly equipped to make sense of the post-September 11th era (Hertsgaard 2002).

International terrorism arguably presents a greater challenge to the existing global political order than the activities of so-called 'rogue states' such as Iraq or North Korea. The Al-Qaeda terror network provides a real problem to modern geopolitical writers, who have tended to work under the assumption that the world can be understood with reference to nation-states and their distinct territories. Extending the 'war on terror' to encompass a range of national/regional problems such as Israel–Palestine and India–Pakistan merely disguises the fact that Al-Qaeda has created a crisis of the geopolitical imagination. The simplistic labelling of global political space (such as Core v. Non-Integrating Gap and/or Wild v. Tame) reflects perhaps an unstated desire to bring some kind of territorial order back to analyses of global political space. As we shall see in Chapter 2, more critical forms of geopolitics seek to reflect and challenge this approach.

Key questions

- How have various US political commentators sought to reimagine the world after the Cold War?
- Why has the United States been accused of being 'imperial'?
- Has terrorism really replaced communism as the most important threat facing Northern countries such as the United States and the European Union bloc?
- Has the military side of world politics increased in salience compared to the Cold War period (1945–1989/90)?

Further reading

Good introductions to geopolitics include G. Dijink, *National Identity and Geopolitical Visions* (London, Routledge, 1996) and J. Agnew, *Geopolitics* (London, Routledge, 2002). Journals such as *Geopolitics* and *Political Geography* should also be consulted. Amnesty International *Rights at Risk* (London, Amnesty International, 2002), G. Simpson, *Great Powers and Outlaw States* (Cambridge, Cambridge University Press, 2003) and K. Booth and T. Dunne (eds.), *Worlds in Collision* (Basingstoke, Palgrave, 2002) provide important sources of information about the pre- and post-September 11th world. Also useful are T. Barnett, 'The Pentagon's new map', *Esquire* (Mar. 2003): 1–5, R. Kagan, 'Power and weakness', *Policy Review* 113 (2002): 1–18 and R. Kagan, *Paradise and Power: America and Europe in the New World Order* (New York, Monthly Books, 2003). Important critiques of recent writings on imperial America and the 'clash of civilisations' include E. Said, 'The clash of ignorance', *The Nation* (22 Oct. 2001): 9–12 and M. Cox, 'The new liberal empire', *Irish Studies of International Affairs* 12 (2002): 39–56. On the fragility of American power see I. Wallerstein *The Decline of American Power* (New York, New Press, 2003) and M. Mann *Incoherent Empire* (New York, Verso, 2003).

Chapter 2

THE NATURE OF GEOPOLITICS AND GLOBALIZATION

Key issues

- What is globalization?
- How do realist, idealist and critical geopolitical approaches to world politics differ from one another?
- What are the differences between strong and weak forms of globalization?

Globalization has become *de rigueur* within popular, political and academic circles (Held *et al.* 1999, Scholte 2000, Waters 2001, Steger 2003 and see Glossary). In the business and management world, it signifies an apparently 'borderless world' in which trade, commerce and money can enjoy unimpeded movement over space and through time. For some government leaders and political scientists, globalization conjures up a world in which the interstate system and state sovereignty have been challenged, perhaps even fatally undermined, by the **transboundary** flows of people, ideas, commodities, finance, disease, drugs and even terror (McGrew 2004). The spread of the SARS virus from one continent to another in 2003 highlighted once more the capacity of dense airline networks to contribute to rapid diffusion of disease. In a more popular cultural vein, these processes and flows appear to signal the emergence of a 'McWorld' based on the widespread adoption and consumption of particular products such as McDonald's hamburgers, Nike shoes and Levi jeans and, in some cases, expressions of local resistance to such domineering trends (Barber 1996, Friedman 2002). In short, globalization indicates a 'power-shift' in the lives and conditions of citizens and our world based on states and clearly defined national territories (Held 2004).

There are also, however, those who are cautious if not actually dismissive of the 'transformative' powers associated with globalization (see McGrew 2004 for a review). Some commentators argue that the phenomenon of globalization is exaggerated, not least because supporters tend to over-emphasize the demise of the nation-state and territorial boundaries (e.g. Hirst and Thompson 1996). From a 'Southern' perspective, globalization often appears to refer to processes and flows that benefit highly industrial Northern democracies at the expense of regions such as Central America and Sub-Saharan Africa (see Chapter 3). For many people simply struggling to survive on a daily basis (often with limited mobility), the world of the Internet, the mobile phone and the 'global advertising' must seem incongruous. This has prompted a current of critical opinion which has argued that deepening inequalities need to be given due consideration in any study of globalization and its impact on politics, cultures and economies.

It is therefore expedient to review the most significant approaches to geopolitics alongside those concerned with world politics. At the same time we shall consider how globalization might influence those theories and their findings. The field of critical geopolitics is considered first, followed by a section on the body of political thought called **political realism**, which is highly popular within academic and policy-making circles especially within the United States. It is an approach to world politics that bears similarities with traditional geopolitical approaches considered in Chapter 1. **Liberal** approaches are explored next, because they seek to combine liberal and elements of realist ideas on world politics. Finally, we turn to some of the recent **critical geopolitical** attempts to reconceptualise geopolitics, global politics and globalization. While this chapter does not consider in detail bodies of thought such as world-systems theories and feminist approaches to politics, it nevertheless aims to demonstrate that a variety of approaches can be brought to bear on world politics. Subsequent chapters draw upon these other literatures in order to demonstrate that there are no right or wrong perspectives because each depends upon specific conceptions of the 'political' and the 'geographical'.

Theories and geopolitics

In a world where there is much to know, there are also many ways of knowing. Claims to one particular way of knowing have frequently been exposed as either misrepresenting or excluding a variety of histories, places and contemporary experiences. Feminist commentators have been highly influential in exposing the fiction that there is a particular place from which one could get a God's eye view of the world (Rose 1993). Thus, the claim of the modern geopolitical writer to survey the world independently of ideology or prejudice would be considered intellectually dubious as well as arrogant. This does not mean, however, that we are unable to make any kind of meaningful statement about the world around us. Rather it implies that we need to consider very carefully the ontological and epistemological basis from which we make our claims about the world.

Explaining contemporary world politics is extremely complex, not only because the range of materials available is substantial but also because the scope of interpretation is wide-ranging. Within the social sciences and humanities, it is now generally accepted that all forms of explanations are, in some sense, theoretically based. For the purpose of this chapter, it is assumed that this theoretical corpus refers to an acceptance of a particular subject matter, which enables choices to be made between which issues or facts matter and which do not. Hence, theory is not considered to be an option just because we are ignorant of the source from which a view of the world has been inherited.

As Robert Cox once noted, 'theory is always for someone and for some purpose' (Cox 1981, cited in Gamble and Payne 1996: 6). The challenge for students of geopolitics and world politics is, amongst other things, to seek to be as *explicit* as possible regarding these theoretical assumptions about the world.

It is only comparatively recently that formal academic disciplines such as International Relations (IR) have been established in universities and institutions. IR was created in 1919 within the Department of International Politics at the University of Wales at Aberystwyth. One of the founders of this discipline, David Davies, argued that IR would help to prevent future wars because the scientific study of world politics would highlight the causes of political problems and in doing so would contribute to the peaceful resolution of global tensions. In the immediate aftermath of the First World War, IR scholars devoted much energy and time to investigations of new forms of conflict mediation and the promotion of new international institutions such as the League of Nations. Labelled idealists by their opponents, proponents of this approach developed a powerful normative element – a concern to promote a model for the world rather than a commentary on the actual condition of world politics. In contrast, the earliest writers on geopolitics were more concerned with the interaction of states and territories rather than with attempts to improve the condition of the world. (There have been honourable exceptions, such as the American geographer Isaiah Bowman and his idealistic geopolitical text on the world after the First World War (see Bowman 1921).) While there were important differences between some of the earliest writers on international politics and geopolitics, there was nevertheless a common approach, which dominated the present century, namely political realism. The intellectual and political contribution of realism will be compared to other geopolitical approaches to world politics such as liberal institutionalism and globalization.

The crux of this chapter is that geopolitics and IR as academic fields are composed and shaped by the interplay of the real world and various fields of knowledge. Understanding the political world depends to a great extent on how we define that world in the first place. A point which will be reiterated is that there is no one overwhelming consensus concerning geopolitics. Geopolitics is an intellectually contested field. Some commentators concentrate on the geographical significance of the nation-state and the international system while others focus on the globalization of world politics. In the spirit of intellectual openness, I should declare for the record that my sympathies lie with 'critical geopolitics' and as such I am critical of much of the academic legacy of traditional geopolitics, which has tended to be preoccupied with geographical problem solving and proffering policy advice to national governments (see below). I also believe that the institutions of the nation-state, while not being overwhelmed by transboundary flows and processes, nonetheless have had to address and confront an increasingly unruly world. The capacities of states, however, clearly vary and inequalities of power, access and influence need to be acknowledged. What is important, regardless of which particular understandings one adopts in attempting to interpret global political spaces, is the recognition that these positions will have implications for understanding and explanation.

Traditional and critical geopolitics

The Swedish political geographer Rudolph Kjellen first used the term 'geopolitics' at the end of the nineteenth century (see Dodds and Atkinson 2000). Since its formal inception, geopolitics has enjoyed a contested and controversial intellectual history. Kjellen's definition of geopolitics as 'the science, which conceives of the state as a geographical organism or as a phenomenon in space' found favour in interwar Germany (cited in Parker 1985: 55). Accused of being closely associated with the expansionist politics of Nazi Germany, it was condemned as poisonous by some post-1945 Anglo-American political geographers. Despite the criticism, geopolitics has been a 'travelling theory' *par excellence* in the sense that it has entered a wide variety of disciplines and geographical regions including Latin America, India and Japan. Academic work on geopolitics has often been conflicting, contradictory and confusing because of the variety of approaches brought to the historical examination of this intellectual field and contemporary analyses of world politics (O Tuathail and Dalby 1998, Agnew, Mitchell and O Tuathail 2003).

In the second edition of *Geopolitics*, John Agnew argues that geopolitics at the turn of the present century was inspired by a *particular way of viewing the world* (Agnew 2002). The invention of the term 'geopolitics' coincided with a certain modernist belief that it was possible to view the world in its totality. The earliest texts of geopolitics reflected the belief that the European observer possessed the necessary intellectual and conceptual framework for viewing the world as an external and independent 'object'. The earliest innovators of geopolitics in Europe and America such as Halford Mackinder and Nicholas Spykman tended to view geopolitics as a form of geographical reasoning, which stressed the capacity of states to act within a changing global arena. Geopolitics was, therefore, a decidedly state-centric enterprise in the sense that the nation-state was paramount and geopolitical writers were eager to offer policy advice. Moreover, the physical environment was frequently conceptualized as a fixed stage on which political events occurred rather than a dynamic and shifting problem which influenced the nature of world politics.

The major difference between **traditional geopolitics** and the more critical approaches is that the latter promote an opening up of political geography to methodological and conceptual re-evaluation (see Table 2.1).

Composed of various strands of social theory, critical geopolitics has sought to consider the ways in which geographical discourses, practices and perspectives have measured, described and assessed the world. The inspiration for critical geopolitics lay in a belief that traditional political geography had failed to disrupt the widespread 'depoliticization' of human geography in the 1950s and 1960s. Boundary studies, for instance, were concerned more with the function and typology of frontiers than with the provision of a critical evaluation of their significance within the international system. Boundaries are central to the discourse of sovereignty as they provide *inter alia* the means for a physical and cultural separation of one sovereign state from another. More recently, the pioneering work of the French geographer Yves Lacoste made a valuable contribution to an agenda which focused on the role geographical knowledge plays in consolidating military power and state-centric politics.

Table 2.1   Traditional and critical geopolitics

Traditional geopolitics	Critical geopolitics
National sovereignty	Globalization
Fixed territories	Symbolic boundaries
Statecraft	Networks/interdependence
Territorial enemies	Deterritorialized dangers
Geopolitical blocs	Virtual environments
Physical/earthly environments	
Cartography and maps	Geographic information systems (GIS)

Source: Adapted from O Tuathail and Dalby 1998

Yves Lacoste was Professor of Geography at the University of Vincennes in France during the 1960s and 1970s. His unhappiness with the academic and political state of geography prompted him and his colleagues to create a new journal called *Herodote* in January 1976. In a famous study entitled *La Géographie, ça sert, d'abord, à faire la guerre!* (Geography is first and foremost for the waging of war), he argued that geographical knowledge had contributed to military power and state-centric politics. In his analysis of the American bombing of dikes in the Red River in North Vietnam, Lacoste demonstrated that the geographical information gathered on the region was being used by American forces to target and destroy the food-growing potential of the region (Lacoste 1973, 1976). Political geographers were reminded that the relationship between geographical knowledge and political power could actually be violent. Lacoste broke with the apolitical aspects of French political geography and clearly stated his belief that geographical work should be located within ongoing political struggles and concerns. Later, Lacoste used *Herodote* to raise a series of issues such as decolonization, immigration, Islamic politics and nuclear missiles. Many of his ideas concerning the role of geographical knowledge in informing foreign policy and military politics have been drawn upon within Anglophone critical geopolitics (Hepple 2000).

Geopolitics is also no longer considered to be the study of statecraft and the Great Powers (the management of international affairs and the ideas that have influenced the practices of diplomacy). Instead it is now perceived as delineating an intellectual terrain concerned with and influenced by the interaction of geography, knowledge, power and political and social institutions. Critical geopolitical writers have argued that geopolitics is a discourse concerned with the relationship between power-knowledge and social and political relations. The adoption of such a position leads authors such as John Agnew, Gearoid O Tuathail and Simon Dalby to propose that world politics has to be understood on a fundamentally *interpretative* basis rather than in terms of a series of divine 'truths' such as the fundamental division of global politics between land and sea powers. For the critical geopolitical writer, therefore, the really important task is interpreting theories of world politics rather than repeating often ill-defined assumptions and understandings of politics and geography.

Challenging conventional categories of international or global politics is part and parcel of a critical evaluation of the role of geographical knowledge and its influence on social and political practices. The emergence of critical geopolitics and geopolitical economy in the 1980s is an indication of political geography shifting from an empiricist past (which assumes that the facts speak for themselves) towards a theoretically informed field of enquiry (O Tuathail 1996, O Tuathail and Dalby 1998). In alliance with critical theories of international relations, critical geopolitics has sought to develop theories of world politics which acknowledge the ambiguity, contingency and uncertainty of the world we live in. Like other developments within the social sciences and humanities, critical approaches to world politics tend to share the postmodern scepticism that the world can be rationally perceived and interpreted through particular techniques.

The starting point for critical geopolitics is to argue that conventional perspectives on geopolitics and international politics ignore the assumptions that underpin those positions in the first place. Critical thinking poses questions such as how current situations come to exist or how power works to sustain particular contexts. Critical geopolitical writers, in contrast to realist observers, argue that the assumption of a detached and objective researcher recording the observable realities of international politics is fallacious. Far from being objective, the research perspective of realism often contributes to the presentation of a view, which appears to legitimate the power politics of states. In contrast, critical approaches to world politics would suggest that unless one challenges or questions contemporary structures and power relations then academic approaches run the risk of merely condoning existing practices. Critical geopolitical scholars now acknowledge that their approaches to world politics are self-consciously situated within a body of conceptual and methodological assumptions about the world. The theories on world politics are not detached from the world we seek to describe and explain, and by acknowledging this point critical theorists may contribute to the development of practical ideas regarding progressive social and political change and how it can be promoted (see Table 2.2).

The analytical framework of critical geopolitics is derived from a mixture of sources including discourse analysis, international political economy, feminist approaches and postmodern social theory. The greatest influence on the literature of critical geopolitics has been the Foucauldian insistence that one must explore the power-knowledge

Table 2.2 Theories of world politics: key themes

Realism	Liberalism	Critical geopolitics
• National sovereignty	• National sovereignty	• Interdependence
• States	• States and non-state	• Globalization
• Military power	organizations	• Networks and nodal points
• Anarchical world	• Limited international cooperation	• Representations of global space

nexus in discourse. As O Tuathail and Agnew noted, 'Our foundational premise is the contention that geography is a social and historical discourse which is always intimately bound up with questions of politics and ideology . . . geography is a form of power-knowledge itself' (1992: 198). In contrast to earlier writings, geopolitics is now not considered to be a neutral technique or device for viewing the world; instead it is seen as a discourse which can be employed to represent the world in particular ways. Thus, the first and most noteworthy source of critical geopolitics was derived from an investigation of the discourses of geopolitics and international relations. Such a position implies that perceptions of the world are derived from a series of assumptions, rules and conventions that are brought to bear by those seeking to explain events and circumstances.

Edward Said's work on Western representations of the Middle East provides one of the clearest examples of how Foucault's insights on discourse and genealogies of power-knowledge are suited to research. Said's famous book, *Orientalism*, charts the creation and evolution of a series of imaginary geographies which constructs the Middle East within Western geographical imaginations. Using British, French and American literary sources on the Middle East (Orient), he argues that the Middle East's complex place and cultural characteristics were reduced to a few defining features such as the 'threat' posed to the Euro-American culture by Arabs and the Muslim faith. *Orientalism*'s concern for the 'distribution of geopolitical awareness into aesthetic, scholarly, economic, sociological, historical and philosophical texts' (1978: 12) has been extremely thought-provoking.

Discourses are seen to influence the rules and conventions by which political behaviour is structured, regulated and judged. Critical geopolitics has argued that discourses play a prominent role in mobilizing certain simple geographical understandings about the world, which assist in the justification of particular policy decisions. Political speeches, for instance, offer possibilities for the promotion of certain ideas to influential actors in world politics. The use of symbolism, metaphors and tropes can be crucial to the shaping of political understandings of specific circumstances. The example given in 'In focus 2.1' may help to elucidate this point.

Critical geopolitics argues that geopolitics should be conceptualized both as a form of **discourse** and as a political practice. Agnew and O Tuathail acknowledged in their investigation of the Cold War the geopolitical reasoning of American political figures such as Ronald Reagan:

> Political speeches and the like afford us a means of recovering the self-understandings
> of influential actors in world politics. They help us understand the social construction
> of worlds and the role of geographical knowledge in that social construction (O Tuathail
> and Agnew 1992: 191).

Geopolitics reconceptualized as a discourse and a form of political practice has several implications:

1. Geopolitics should be considered as a political activity carried out by a range of political actors and not limited to a small group of academic specialists.
2. Geopolitical reasoning employed by American statesmen during the Cold War points to the fact that 'unremarkable assumptions about place and their particular

In focus 2.1: American representations of Libya as a 'terrorist state'

President Ronald Reagan's decision to label Libya a 'terrorist state' in 1985–6 was an important prelude to a growing reassessment by the United States of the new threat posed by Islamic fundamentalism. The bombing of the Libyan capital, Tripoli, in 1986 was the culmination of an American belief that Libya had funded terrorist activity in the Middle East and Europe. Rather than being simply a piece of rhetoric designed to provoke a political argument, the political outcome was to damage US–Soviet relations just as they appeared to be improving. Widespread public support in the USA for the Tripoli bombing may well have been assisted by sympathetic media coverage and a growing trend in Hollywood to produce films locating threats against America in the Arab world.

In December 1988 a Pan-American airliner carrying several hundred people from London to New York exploded over southern Scotland. After months of investigation, the American and British police named two Libyan airline officials as suspects for this act of terrorism. After 10 years of oil and other forms of sanctions (against Libya) following American pressure on the United Nations, the Libyan leadership agreed that the two Libyan suspects would face criminal proceedings in the Netherlands under Scottish law. One of the suspects was convicted in 2001 and in 2003 the United Nations voted to lift sanctions against Libya after it offered a substantial compensation package to the victims of the Pan-American attack. Libya, however, remains a 'rogue state' in the view of the George W. Bush administration, despite its important decision in December 2003 to desist from further developing WMD. The British prime minister Tony Blair paid a visit to Libya in May 2004 and arguably contributed to the latter's 'diplomatic rehabilitation'.

While the American and UK governments have condemned Libya for supporting terrorist activity in the past, the fact that the Americans carried out an illegal bombing raid on Tripoli in 1986, which also killed innocent civilians, has been ignored.

identities' can be highly significant. The assumed threat of the Soviet Union overwhelming American or Western civilization often drew upon long-standing geographical depictions of the Euro–Asian landmass being populated by Asiatic hordes intent on conquering European and Slavic peoples.

3. The current distribution of power within the international system means that some states such as the USA are in a better position than others to influence the production and circulation of political discourses and thus possess the capacity to shape geopolitical understandings of the world.

4. Critical geopolitics argues that expressions of geographical difference contribute to a politics of identity formation. The widespread depiction of the Soviet Union as an 'evil other' during the Cold War not only helped to secure America's identity as a bastion of liberty and democracy but also empowered successive administrations to target anyone considered to be an internal subversive or crypto-communist.

Critical geopolitics and geopolitical economy

The distribution of power within the international system is a major consideration for the geopolitical economy. The two political geographers most closely associated with the geographical relations of economic and political domination and dependence are John Agnew and Stuart Corbridge (Agnew and Corbridge 1995). Critical geopolitics and geopolitical economy share a number of considerations:

1. States are not the only influential force in international politics. The activities of multinational corporations, non-governmental organizations and firms are considered to be of importance. It is abundantly clear that states have to operate within a world economic system not only where flows of capital and technology transcend territorial boundaries but where the activities of business corporations who operate in more than one country or region also influence this process. Multinational corporations often enjoy considerable independence from particular governments even if they are identified as 'American' or 'Japanese' firms.
2. The presumption that states pursue so-called national interests often underestimates the importance of sectional interests, which may be represented as national interests for political reasons.
3. Analyses of international politics often neglect patterns of economic relations to the detriment of the structure of the international political system. International relations are thus reduced to a concern for the interaction between states through diplomatic and political arenas rather than focusing on the reciprocal action between the world economy and the power of the state.

Geopolitical discourse participates in the construction of geographical significance for places and regions, which can be linked to wider material interests. The capacity of the United States, as the largest economic and military power, to represent the world economy in particular ways (for example, as an open and unrestricted arena of free trade) matters due to its influence on international financial organizations such as the World Bank and the International Monetary Fund (Herod, O Tuathail and Roberts 1998). That does not imply, however, that individual states such as the USA or Japan can determine an increasingly integrated global economy. The strategies employed by Great Powers to maximize specific interests are frequently 'scripted' through particular representations about the world economy and international politics. Geopolitical economy is, therefore, concerned with the interaction between, *inter alia*, Great Powers, the international political sytem and economic processes.

The importance of discourse and representational practices has been a hallmark of critical geopolitics. Many papers and books have explored how foreign policy professionals and academics have depicted and represented global political space. The formal geopolitical reasoning of these individuals can then be seen as contributing to particular visions or vistas of world politics. Said's concern for imaginative geographies and representational practices has also stimulated interest in rethinking the formal disciplinary history of Anglo-American geopolitics. Many writers from Mackinder onwards have argued that the countries and regions of the 'East' have threatened the Western world. Mackinder's earliest geopolitical paper (1904) on 'The geographical pivot of history' identified a 'Heartland' within the Soviet Union and claimed that Western powers would be threatened by Eastern powers: 'Were the Chinese, for instance, organised by the Japanese, to overthrow the Russian empire and conquer its territory, they might constitute the yellow peril to the world's freedom' (1904: 430). In a similar vein, American geopolitical writers in the 1940s and 1950s argued that the menace from the East in the form of the Soviet Union threatened to imperil the free world.

Realism and the 'Westphalian model' of world politics

The international lawyer Richard Falk coined the term the 'Westphalian model' of world politics in order to describe a world allegedly characterised by the territorial sovereignty of states, an anarchic international arena, legal and political equality between states, the inherent right of states to use force in order to settle disputes and limited co-operation between states (Falk 1999). The term 'Westphalian' is derived from the 1648 Treaty of Westphalia and is commonly held to have ushered in the role of European states in shaping international politics. The institution of the nation-state was henceforth considered to be the premier political organization in European and later world politics. More commonly, however, 'Westphalian' is associated with political realism by academics and policy makers.

Political realism (often shortened to realism) is widely recognized as the most influential body of literature associated with IR and geopolitics. This approach to world politics should not be confused with the philosophical realism of Roy Bhaskar or Andrew Sayer. The premise for the earliest realist writers in the 1920s and 1930s was that the so-called idealists had misunderstood the nature of world politics. For the realist the world was unpleasant and populated by generally selfish human beings intent on self-gratification rather than collective improvement. The realist view on human nature is inevitably pessimistic and assumes that international politics can never be comparable to domestic politics. The principal political force remains the nation-state and the major determinant of international relations is the balance of power between states.

Moreover, the interactions of sovereign states occur in an international arena shaped by anarchy rather than peaceful co-operation because self-interested states are not subject to the regulatory authority of any supernational body (world government). According to some realist commentators, the League of Nations (created in 1919–20) was doomed to failure because it did not acknowledge that states were intent on maximizing their national interests at the expense of international co-operation. In the final analysis, realists argue that states will often rely on military force in order to achieve their own ends. Even in the era of the United Nations (created in 1945), the Iraqi invasion of Kuwait in 1991 or the American bombings of Sudan and Afghanistan in 1998 serve to consolidate this basic conclusion.

In contrast to idealism, therefore, realism was concerned with the apparent realities of world politics. However, this supposed concern with the realities of global political life did not guarantee that this approach was either commonsensical or neutral. Like other approaches to world politics, realism makes a series of assumptions about political life, human nature, the international system and the interactions of nation-states. As Rob Walker has demonstrated, realist views of political life embody a fundamental contrast between life inside and outside the state (Walker 1988, 1993). Within the state it is possible to live the 'good life' and to become part of a society characterised by citizenship, community and culture. Outside the state, the notion of an international community of people is effectively abandoned and replaced by an interpretation which judges international relations to be dominated by war, violence, uncertainty and selfishness. The conception of community was not, however, abandoned altogether as realists argued that co-operation between states was possible in international society based on rules, laws and customs which moderated the behaviour of states. The creation of the United Nations in 1945 was widely considered to be a step towards the peaceful regularization of international relations.

Those who believe that the 'Westphalian' model does not capture the complexities of modern political life have nonetheless hotly disputed realist views of world politics. Frequent criticism is levelled at the approach favoured by realists, who tend to be more concerned with the role and behaviour of states, to the detriment of other international bodies such as firms, non-governmental organizations and international organizations whose contributions are often significant in shaping distinctive political agendas. Realists argue that the pursuit of military security is the primary objective of states. Their definition of security has frequently neglected to address other factors such as environmental, cultural and economic forms of security. This concern for the military and political aspects of security has led to the underestimation of social and cultural factors in shaping world politics. Realists tend to view the international system as an anarchical arena populated by states (regardless of their social and cultural backgrounds) which function in an undifferentiated manner. The implication of this position is that realists believe that the nature of the international system is, for instance, unaffected by cultural variation.

Scholars such as Hedley Bull and Rob Walker have argued that the realist depiction of the international system as an anarchical arena depends upon the representation of domestic political life as shaped by order and relative peace. This depiction of the international system reveals more about the power of dichotomization than about the way in which categories such as the national and international interact with one another to produce particular understandings of political life. The challenge for students (according to Walker) is to present an account of the international which is different from, but not a negation of, the national. Finally, the assumption by realists that there is either an independent or a commonsensical way of seeing the world is epistemologically naive. The categories brought to bear on the world by realist analysis have implications for understanding of politics, international society and territorial states. For example, an Islamic understanding of world politics would elicit a different reaction to the problem of international politics than a Western understanding.

In response to these criticisms, realist writers such as Robert Gilpin and Kenneth Waltz have sought to bring a theoretical and conceptual rigour to political realism. Since the 1970s, neo-realism has been a highly significant body of thought in IR, fighting to remove some of the unscientific methods of realism and the anecdotal use of historical and geographical case examples. The inspiration for so-called neo-realism evolved through the changing circumstances of the world economy and politics. It was becoming increasingly obvious that the state was only one institution (rather than the main actor) within international politics and that a concern for interstate relations needed to be located within a broader political framework which included non-state organizations and transnational relations alike.

The publication of Kenneth Waltz's *Theory of International Politics* (1979) was probably the most significant contribution to the development of neo-realism within Anglo-American IR. Using Karl Popper's conception of scientific method, Waltz argued that realism needed to be reformulated as a positive theory intent to produce law-like propositions for the international system. His aim, as Chris Brown noted, was to produce a theory of the international system rather than to account for all aspects of world politics (Brown 1997: 46). In his dense analysis of the international system, Waltz proposed that two possibilities exist for the international system: a hierarchical or anarchical system. The former is considered to be a system composed of different kinds of units organised under a clear line of authority whilst the latter is composed of units which are similiar to one another. Hence, according to Waltz, the international system is (and has been since medieval times) anarchic in the sense that states co-operate with one another as equals in the absence of any form of world government.

However, Waltz's conceptualization of the international system is also conditioned by the existence of Great Powers such as the USA. As with realism, great stress is laid on the role of these powerful states in maintaining order within the global system. As the economic and political hegemonic power, the USA created a basic political and economic framework for the post-war world based on the United Nations, the Bretton Woods agreement and American military power (see In focus 2.2).

Neo-realism places considerable emphasis on the structure of power within the international system and its impact on the prevailing political order. For Waltz, the existence of two superpowers in the 1970s was preferable to three or more powers because of the capacity to impose stability on the global political order. In the process, the superpower confrontation of the Cold War transformed the world into one large strategic arena.

Many neo-realists would be sceptical of the 'globalization of world politics thesis' because, as they would argue, the state remains the principal actor in an anarchic international arena. In spite of changes to the nature of world politics, states in the post-Cold War era remain committed to the pursuit of national interests and remain cautious about co-operating with other states, as was evidenced in recent events such as the US and European reluctance to intervene in the Yugoslav civil wars (1992–5). In spite of the increased domain of co-operation, neo-realists believe that states retain a rational mindset, motivated by self-interest and self-preservation. In contrast to Kuwait and the oil fields of the Middle East, therefore, Bosnia was not considered to be strategically important by the international community even though there was a desire to end the suffering of civilians.

In turn, the critics of neo-realism have pointed out that these approaches to international politics tend to be inherently conservative in terms of theory construction and

---

In focus 2.2: Bretton Woods

In 1944 an important conference was held in the New England town of Bretton Woods. Under the leadership of the Allied powers, the USA and the UK, the future of the world's economic and financial system was discussed. The delegates to the conference agreed to reverse earlier policies of trade protectionism and to promote a new regime based on a stable money-exchange system and freer trade. The value of each country's currency would be determined by the fixed gold value of the US dollar. Bretton Woods helped create the institutional foundations for the post-1945 world economy. The International Monetary Fund (IMF) and the International Bank for Reconstruction and Development (later known as the World Bank) were established as part of the institutional strengthening of the world economy. In 1947 the General Agreement on Tariffs and Trade (GATT) provided the final element of a global operation designed to promote economic recovery after six years of conflict.

The Bretton Woods regime survived until the early 1970s when the US administration abandoned the gold-based fixed-rate system amidst fears that US industries were becoming uncompetitive. Subsequently the world economy was plunged into a period of instability culminating in high unemployment, energy crises and public-sector shortfalls. By the late 1970s, a new generation of right-wing political leaders entered office advocating a 'liberal' approach to international economic affairs.

---

political aspiration. Waltz's theory of the international system simply accepts the existence of anarchy rather than seeking to analyse the ways in which the construction of anarchy facilitates particular interests (Brown 1997: 56). In Robert Cox's terms, neo-realism is a problem-solving set of theories rather than a series of critical theories which seek to change particular situations (Dalby 1991). For critical theorists, neo-realism is an impoverished approach to world politics because it does not concern itself with either human emancipation or the search for alternatives to the present political condition.

The position of political geography within the corpus of realism and neo-realism is therefore difficult to locate because few political geographers have explicitly acknowledged their theoretical assumptions about the international system or politics. By contrast, American IR theorist Hans Morgenthau claimed that, 'International politics, like all politics, is a struggle for power. Whatever the ultimate aims of international politics, power is always the immediate aim' (Morgenthau 1948: 27). Within conventional geopolitics, many writers argue that political life is dominated by the interaction of states in particular geographical settings. No political geographical writer compares to the status of Morgenthau, whose best-selling book *Politics Among Nations* listed the six major principles of political realism (Morgenthau 1948). It has been argued that the implicit assumptions of traditional political geography have been inspired by realist thought: 'As it informs a rather large and influential literature on geopolitics and military affairs, for example, realism has often degenerated into little more than an apolitical apology for cynicism and physical force' (Walker 1993: 107). Peter Taylor has argued that traditional geopolitical thinking in the mode of Mackinder and Mahan was inspired by a tradition of power politics within international relations (Taylor 1993). Mackinder's model of competing land and sea powers was inspired by his commitment to promoting British imperial interests in the face of overseas competition from Germany and Russia. The development of railways was considered crucial to the balance of power between imperial nations because it would allow traditional land-based powers such as Russia to control vast land areas through speed of travel. The identification of the Euroasian landmass as a 'geographical pivot of history' pointed to the geopolitical significance of particular territories in the struggle for control over the earth's surface.

Traditional geopolitics has also been underwritten by many of the assumptions of political realism concerning the nature of the international arena and the significance of state sovereignty and national interests. In contrast to realist analysis of international politics, however, political geography and geopolitics have focused on the power of the land and the sea to shape international relations. Classical geopolitical writers such as Mackinder endowed the 'Heartland' with the potential to influence world politics at the expense of the so-called rimlands and outer crescents. Fixed assumptions about the geographical significance of places littered the geopolitical discourses of European and American political geographers. Geographical divisions were considered timeless and thus immune to human alteration. As the American political scientist Ladis Kristof once argued: 'The modern geopolitician does not look at the world map in order to find out what nature compels us to do but what nature advises us to do, given our preferences' (Kristof 1960: 19). The capacity of human observers to influence understandings of world politics was diminished when the meaning of place and region was considered static rather than capable of change. Geography was reduced to the role of simply providing a territorial stage on which the interactions of states unfolded. Recent work within political geography has suggested that this is a very restricted view of geography which ignores how and why geographic spaces and places are made significant through the processes of discursive construction.

In conclusion, realism has frequently been condemned for being an incomplete intellectual and political project. While it could be argued that an approach which stresses the significance of states, war and national interest is admirable, realist presumptions about the interstate system and national behaviour do not account for many features of world politics. If, for example, the national interest of states is the primary concern for political leaders, why do Nordic countries such as Sweden give substantial amounts of their GDP to the cause of humanitarian and developmental projects in the Third World? This should not imply that the state and state sovereignty are exhausted either as concepts or as political and legal powers, even though many authors have suggested that transnational flows and processes progressively blur national boundaries and identities. Writers sympathetic to liberal approaches to world politics counter that realism fails to explain how the international system constrains and influences state behaviour through a series of conventions, treaties and international organizations such as the United Nations.

Liberalism and the 'UN charter model' of world politics

Richard Falk also coined the term the 'UN charter model' of world politics to describe a world in which: states co-existed with other social and political actors, co-operation was not limited between states, rules and regulations were used to eliminate unacceptable features of world politics such as genocide and war and where the territorial boundaries of states were blurred by transnational and supranational relationships. These series of assumptions are the foundation of the approach to world politics called **liberal institutionalism**. This is an intellectual compromise between liberalism and realism because while it is recognized that states and national interests are important features of the international system, it is proposed that a variety of others also share global political spaces such as the United Nations, intergovernmental organizations and NGOs (Fig. 2.1).

Liberal institutionalism contends that the international arena is not entirely anarchical. Although they would agree with the realist that the sovereign state is the major organization within the international system they would not necessarily accept that there are no checks or balances on the behaviour of states. A series of conflict-mitigating factors and transnational institutions ensure that states do not behave in a selfish and violent manner. These include a variety of intergovernmental and transnational regimes such as the 1959 Antarctic Treaty, which ensured that the polar continent

Figure 2.1 United Nations Security Council
Photo: UN/DPI Photo

Figure 2.2   Antarctica: a zone of peace and co-operation
Photo: Klaus Dodds

has remained a zone of peace and a place for international scientific co-operation (see Fig. 2.2).

The success of the Antarctic Treaty System is undoubtedly based on the fact that 44 states agreed to temper their own national ambitions for the sake of peaceful co-operation and the environmental protection of the region.

Figure 2.3   Kofi Annan, the seventh secretary general of the United Nations
Photo: UN/DPI Photo by Sergey Bermeniev

The most important intergovernmental organization which seeks to promote international co-operation and peaceful exchange is the United Nations (UN). Under the 1945 Treaty of San Francisco, the international community created the UN in the hope that the anguish of the Second World War could be replaced by peace, dialogue and universal solidarity. The purpose of the UN was spelt out in the UN Charter (111 articles), which defined common goals for the world community such as the implementation of particular moral values and standards for international relations. Signatories to the UN Charter had to commit themselves to: the peaceful resolution of disputes, the sovereign equality of all members, the principle of collective security and a range of other social, political and cultural concerns (see Whitaker 1997). The UN sought to maintain order and codify certain forms of behaviour as either acceptable or unacceptable. The UN Charter also established the following bodies: the General Assembly, the Economic and Social Council, the Trusteeship Council, the International Court of Justice and the Secretariat (Fig. 2.3 and Table 2.3).

Table 2.3 Secretary generals of the United Nations

Years in office	Name	Country
1946–1953	Trygve Lie	Norway
1953–1961	Dag Hammarskjöld	Sweden
1961–1972	U Thant	Burma (now Myanmar)
1972–1981	Kurt Waldheim	Austria
1981–1991	Javier Pérez de Cuéllar	Peru
1991–1997	Boutros Boutros-Ghali	Egypt
1997–	Kofi Annan	Ghana

For critics of liberal institutionalism, the performance of the UN is indicative of the difficulties inherent in this body of thought. During the Cold War, the role of the UN was effectively neutralized by a number of 'Great Powers' (China, France, the UK, the USA and the USSR) who made up the permanent members of the Security Council. Armed with the power of veto, these states habitually paralysed the UN and its executive orders, often on the basis that particular UN operations or directives would interfere with their own strategic or political goals. The alleged sovereign equality of UN member states was frequently exposed as 'hollow' during the Cold War as and when the Great Powers either ignored UN resolutions or violated the sovereign rights of Third World states. Although the American invasion of the Dominican Republic in 1965 was declared illegal by a UN resolution, this did not deter the US from pursuing its own strategic objectives in the Caribbean.

In other cases, key strategic allies of the Great Powers such as Israel were allowed to marginalize significant resolutions such as Number 242, which called for a 'just and lasting peace' in the Middle East after the 1967 Arab-Israeli War. As part of this peace process, Israel was supposed to withdraw from the so-called West Bank territory of Jordan, but it resolutely refused to implement this part of the UN resolution.

The UN Charter's opening preamble also invokes 'We the people' while at the same time denying membership of the UN to non-state organizations and stateless peoples such as the Kurdish people in the Middle East. Many scholars have argued that an understanding of world politics can only be achieved by recognizing that 'political life' is not dominated by nation-states. While liberal institutionalists recognize that NGOs, intergovernmental organizations and multinational corporations have significant roles to play, they still tend to overemphasize the role and scope of the state in their accounts of world politics.

Geopolitics and globalization of world politics?

Globalization has emerged as a central point of theorization and debate within the humanities and social sciences. Since the 1990s, **internationalization** has been replaced by globalization because this is considered to be more helpful in analysing cross-boundary interaction. Considerable debate concerning the geographical scope, historical relevance and technological intensity of cultural, economic and political forms of globalization ensued (see Waters 2001, Scholte 2000, Steger 2003). Against such a backdrop, it is unsurprising that contemporary thinking about human affairs has acknowledged that we are all participants in a world of global connections (Walker 1988). The evidence for and against globalization is, however, manifestly disputed by the wide-ranging debate over the origins and significance of globalization (Robertson 1992, Held 1995, Hirst and Thompson 1996). For the supporters of globalization, recent changes in the world system are so profound that global politics, economics and culture have been radically altered. For the sceptics, however, the features ascribed to globalization are either exaggerated or insufficiently located within a longer historical process of world capitalist development. The sceptics conclude that a more careful analysis would reveal that the present levels of integration, interdependence and involvement of national economies and polities are not unprecedented.

According to the sociological writer Roland Robertson, globalization can be understood as a process whereby social relations acquire relatively distanceless and borderless qualities because the world is becoming a single and highly integrated place (Robertson 1992). He argues that there has been an active process of social system building at the global level for at least the last century and a half. The development of international trade and political co-operation has facilitated this evolution. Over time, the global system has become more complex and interdependent because of time–space compression and the development of global consciousness. The former has enabled the creation of more intense interdependencies with the result that sudden changes in one part of the world can have implications for others. Around the world, for instance, the rapid popularity of the Nike running shoe in North America and Europe led to an increased demand in production, which in turn had implications for the workers who produced these shoes in South East Asia. Other examples can be drawn from the environmental sphere, where unregulated industrial development and practices can adversely affect areas in other regions. The uncontrolled burning of the Indonesian forest in September 1997 forced citizens in Singapore to wear protective masks in order to avoid breathing noxious air and caused international air traffic to be diverted.

The development of a global consciousness is related to time–space compression. Robertson argues that global consciousness has been facilitiated by developments in media communications, which allow people to participate in global discourses on 'world peace', 'environmental protection' and/or 'human rights'. Since the 1960s, these sociological processes relating to globalization have intensified around the world. As the French social theorist Paul Virilio once noted:

> And yet critical space, and critical expanse, are now everywhere, due to the acceleration
> of communications tools that obliterate the Atlantic (Concorde), reduce France to a square
> one and a half hours across (Airbus) or gain time with the Train de Grande Vitesse, the
> various advertising slogans signalling perfectly the shrinking of geophysical space of which
> we are the beneficiaries but also, sometimes, the unwitting victims (Virilio 1997: 9).

Places and peoples are being drawn together into the socio-political space of others. This transformation has eroded the principle of state sovereignty in the sense that states and societies are experiencing greater difficulties than ever before in controlling their own affairs within their national territories. For the supporters of 'strong globalization', world politics has been or is being fundamentally changed.

Arguments in favour of 'strong globalization'

The 'evidence' for such a proposition lies in a number of directions and includes the following:

1. Economic transformations in the world economy have meant that national states are losing the capacity to control their own national economies. Interest-rate changes in one economic region swiftly impact upon other regional components of the world economy. Currencies and commodities appear to travel across borders with very little interference from financial institutions and/or states. Within this apparently border-less world, business gurus such as Kenneth Ohmae have argued that the nation-state is an outmoded institution, which is ill-equipped to deal with world markets and borderless transnational corporations (Ohmae 1990). The currency crises in Russia, Brazil and South East Asia would seem to confirm this observation, as states struggled to bolster their collapsing currencies in the midst of recession in 1998–9.

2. Information and communication technologies have promoted the growth of global electronic networks, which enable information to be sent rapidly across the world. The development of the Internet in the 1980s is probably the most significant illustration of the global network society (Castells 1996). From a political perspective, nation-states can often find it hard to control the flow of sensitive information. While information or images censored in one place are often available in another, this has led to the rise of deeply undesirable activities such as political extremism in the form of Neo-Nazism and far-right politics.

3. We live in a global risk society, which has to confront transboundary health challenges such as AIDS and SARS, and other issues such as pollution, which are often beyond the control of either one or a group of states. The movement of people via plane routes (from China to North America, for example) unquestionably diffused the outbreak of SARS in 2003.

Written in the aftermath of the 1986 Chernobyl nuclear disaster, Ulrich Beck's first English-language book on the risk society was a powerful account of how modern societies experience rapid and accelerating change in a host of fields, including informational technologies and financial markets (Beck 1992). Beck's description of these changes identified 'risk' as central to our late-modern culture because so much of our thinking is of a 'what if' kind in the face of uncertain futures. Unsurprisingly, relatively affluent people in the USA and Europe are now spending more money than ever before on insurance policies and US administrations constantly warn about the dangers posed by risks such as nuclear proliferation and 'outlaw states' such as Iraq.

4. The expanding influence of regional organizations such as the EU is a reflection of a growing belief that neighbouring states have to co-operate with one another in order to secure the best possible position within the global political economy (Fig. 2.4). The

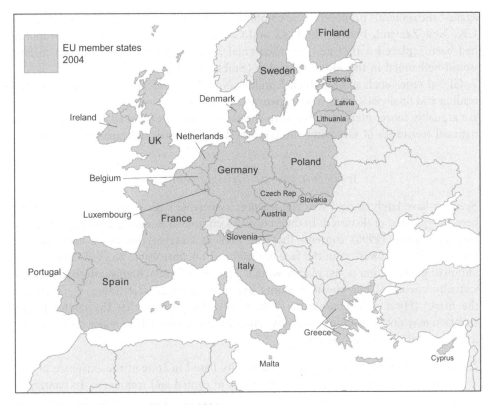

Figure 2.4 The European Union in 2004

European Council of Ministers is challenging the decision-making powers of the member nation-states. The UK government throughout the 1980s and 1990s was instructed by the European Council of Ministers to carry out certain initiatives (such as the culling of BSE-infected cattle herds in the mid-1990s) even though the British Parliament had voted against their full implementation.

5. The rise of transnational corporations (TNCs) means that new forms of global politics are challenging old forms of international politics. In short, states have to compete with a variety of non-state organizations and this has often been perceived as detrimental to sovereign state power. The activities of the 200 largest transnational corporations constitute 50 per cent of the world's total industrial output. The largest companies dwarf the economies of the Global South and are able to seek out cheap economic production sites in a largely deregulated global market.

6. Armed with sympathetic television coverage, NGOs such as Greenpeace have been highly effective in challenging the decision-making powers of governments. One clear example of this capacity to contest governmental strategies was evident in 1989 when Greenpeace in alliance with other environmental NGOs launched a global campaign against proposals to devise a minerals agreement for the Antarctic. Media campaigns

against the minerals proposals were combined with mass public action in Australia, the UK, New Zealand, France, the USA and Canada. Within two years, these proposals had been replaced with a new environmental framework which stressed that mining would be banned in the Antarctic region (Stokke and Vidas 1996).

7. Global cities such as Frankfurt, London, New York and Tokyo are part of a 24/7 trading and financial system which spans the world. As sites of cross-border flows, they are arguably more interconnected with a global circuit of capital than they are tied to national territories of Germany, the UK, the United States and Japan respectively.

Arguments against 'strong globalization'

Sceptics have lately argued that the features associated with globalization have either been exaggerated or distorted. In their powerful critique of economic globalization, Hirst and Thompson (1996) contend that the present state of the world economy has not made states powerless in the face of rapid and uncontrollable flows of capital: 'The notion of globalization is just plainly wrong. The idea of a new, highly-internationalised, virtually uncontrollable global economy based on world market forces . . . is wide of the mark' (Hirst and Thompson 1996: 47–8). They argue that the contemporary situation may not be as unique as many critics have suggested. Five major claims underline their critique of globalization:

1. Economic activity continues to be nationally based in spite of the existence of the world economy and transnational flows of capital and commerce. In contrast to the suggestion that transnational corporations are dominating world trade patterns, it has been shown that they retain on average two-thirds of their assets in the home base and remain embedded in a particular national context. Within the major Northern economies, international business remains closely tied to home territory in the sense of overall business activity, location of sales, declared profits and research and development.
2. Globalization is no more than the sum total of international flows of trade and capital between countries and not an economic system articulated on a global scale.
3. Flows of trade and capital remain overwhelmingly concentrated in self-contained regional groupings such as the EU, North America and East Asia. Contemporary flows of trade and commerce are not, therefore, overwhelmingly global. The geographical reach of world capitalism has actually receded in terms of foreign capital flows and world trade, in the main at the expense of Sub-Saharan Africa and Latin America. Likewise global media networks, for example, are controlled by a limited number of corporations based in the North.
4. National economic regulation is still possible because there is scope for specialization and regulation through institutions and agreements. Pressure from above and below the nation-state in the form of regional economic blocs (e.g. the EU) alerts us to the fact that national regulation has to co-exist with other flows and forces.

5. Most TNCs are not global in the sense that their headquarters and major trading activities are concentrated in the three major trading regions (Europe, North America and East Asia) of the world economy. According to the index of industralization, only 21 out of the top 100 TNCs have a high global profile.

In support of these assertions, Hirst and Thompson suggest that the current trends in the global economy are actually similar to the period between 1870 and 1914, and that arguments pertaining to the uniqueness of the present world economy are overstated. States have had to deal with processes such as internationalization for a considerable time. Moreover, the idea of transnational corporations taking over significant decision-making powers of states overestimates the degree to which TNCs are actually independent from the affairs of state. Most TNCs are national companies which happen to trade internationally, and this pattern of trading is still concentrated in particular regions of the world economy such as East Asia, the EU and North America. Globalization, under this analysis, is found to be a phenomenon that is overwhelmingly Northern rather than global in the sense of the geographical distribution of foreign direct investment, communication networks and trading patterns.

Hirst and Thompson's critique of globalization is based on a series of observations on the world economy grounded in a quantitative evaluation of trade flows, business activities and international politics. However, their thesis fails to deal adequately not only with the qualitative shifts in the nature of global exchange and interconnections but also with the substantial constraints on national decision-making. David Held has argued that there has been a considerable shift in the nature and extent of global interaction as compared to the nineteenth century:

> For there is a fundamental difference between, on the one hand, the development of particular trade routes, or select military and naval operations or even the global reach of nineteenth century empires, and, on the other hand, an international order involving the conjuncture of: dense networks of regional and global economic relations which stretch beyond the control of any single state . . . extensive webs of transnational relations and instantaneous electronic communications . . . a vast array of international regimes and organizations which can limit the scope for actions of the most powerful states; and the development of a global military order (Held 1995: 20).

The consequence of such a shift in the global order is that the nation–states and national economies have to co-exist with a range of networks and social actors. International politics became more complicated in the late twentieth century as multinational corporations (accounting for two-thirds of the world's trade) contributed to the development and intensification of global circuits of production and exchange.

As a counter to these arguments concerning either the triumph of global capitalism and/or the calling into question of globalization itself, this author would argue that globalization is probably best considered as an intensification of interaction between national and transnational social formations operating through the interstate system. Owing to external influences, this means that the state has lost some capacity to regulate a national economy through deregulation, exchange and interest rates and fiscal policy. Increasingly, Northern states such as the United Kingdom and Germany have developed macro-economic and political policies which promote global competitiveness and encourage inward investment but remain susceptible to externally influenced interest rates and currency fluctuations. From a cultural perspective, it is abundantly clear that the world has not been reduced to a homogenous cultural mass. Cultural life has been affected by vast flows of people, business and tourism. Traditional boundaries between territorial and social spaces have become blurred, and hybrid cultures and identities have been the defining feature of this interweaving of the local and the distant. As Anthony Giddens has argued:

> Globalization is not just an 'out there' phenomenon. It refers not only to the emergence
> of large-scale world systems, but to transformations in the very texture of everyday life.
> It is an 'in here' phenomenon, affecting even intimacies of personal identity . . .
> Globalization invades local contexts of action but does not destroy them; on the
> contrary, new forms of local cultural identity and self expression are causally bound
> up with globalizing processes (Giddens 1996: 367–8).

From a political perspective, various writers have sought to convey a sense that transnational governance challenges state sovereignty over domestic affairs and the international system based on interstate diplomacy. In contrast to Hirst and Thompson, it has been argued that new forms of governance have emerged on the world stage which include: governments and firms negotiating amongst themselves, transnational structures such as the UN and the growing influence of non-governmental organizations within political spheres (see below). Other commentators have pointed to the growth of a global civil society where international social movements and the mass media contribute to a new civic awareness of human tragedies, environmental disasters, pollution, war and structural inequalities (Shaw 1996). The end result of these kinds of fora is not the eradication of the state and its power to regulate a national economy but a reworking of national economic and political life in the context of transnational flows of capital, commerce and governance.

Conclusions

This chapter has been devoted to the notion that there is no one particular intellectual pathway for the comprehensive study of world politics. While this account has been critical of realist and neo-realist accounts of international politics, it should be recognized that the pursuit of national security is a major issue for some states and regions such as Palestine, Israel, Lebanon and the wider Middle Eastern region. It is a major preoccupation of contemporary American foreign policy following the September 11th attacks. Although recent interest in the so-called globalization of world politics has drawn attention to these interrelationships, critics of globalization have warned that some writers have overemphasized the declining power of the state and underestimated the fundamental differences which exist between North and

South. However, other theorists argue that realism does not really consider *transnational* relations between states and non-state actors. As a result, it presents a rather restricted view of world politics, which fails to acknowledge that there exists a diffusion of networks and actors including NGOs, IGOs and multinational corporations.

Key questions

- Why are realists pessimistic about human nature?
- Can bodies such as the United Nations prevent international politics from being anarchic?
- How and why does critical geopolitics differ from traditional geopolitics?
- Why does globalization pose considerable challenges to the interstate system based on discrete sovereignty and territorial jurisdiction?

Further reading

For good introductions to globalization and global politics, see D. Held *et al.*, *Global Transformations* (Cambridge, Polity, 1999), J. Baylis and S. Smith (eds.), *The Globalization of World Politics* (Oxford, Oxford University Press, 2001), K. Booth and T. Dunne, *Worlds in Collision?* (Basingstoke, Palgrave, 2002) and earlier writings such as C. Brown, *Understanding International Relations* (Basingstoke, Macmillan, 1997). On critical geopolitics see G. O Tuathail, *Critical Geopolitics* (London, Routledge, 1996) and G. O Tuathail and S. Dalby (eds.), *Rethinking Geopolitics* (London, Routledge, 1998). For a flavour of realist and liberal institutionalist work on international politics see H. Morgenthau, *Politics Among Nations* (New York, Alfred Knopf, 1948), K. Waltz, *Theory of International Politics* (Reading, Addison Wesley, 1979) and R. Keohane and J. Nye, *Power and Interdependence* (Cambridge, MA, Harvard University Press, 1989).

Chapter 3

GLOBAL APARTHEID AND NORTH–SOUTH RELATIONS

Key issues

- Is globalization fundamentally predicated on a form of spatial apartheid?
- What role did the Third World play during the Cold War? How did Third World states seek to resist Cold War divisions?
- Did the ending of the Cold War lead to a radical shift in North–South relations?
- What factors have shaped US–Latin American relations in the post-Cold War era?

The collapse of the Cold War (1989 onwards) focused attention once more on the structure of the global political economy and the possibilities of ameliorating divisions of wealth between North and South. The 1990s demonstrated, however, that these divisions between rich and poor are worsening and some of the most extreme pockets of poverty are now to be found within the former Soviet Union in places such as Armenia and Azerbaijan (Bradshaw and Stenning 2004). The *Human Development Report* published by the United Nations in 2003 makes for depressing reading: over 50 countries witnessed drops in national income over the 1990s, 30,000 children continue to die daily from preventable illnesses, the richest 1 per cent of the world's population now receives as much income as the poorest 57 per cent. Twenty–eight million people are thought to have contracted AIDS in Africa and over 13 million children died of diarrhoea (an utterly preventable illness which would reduce dramatically if access to clean water were secured) in the 1990s. Even allowing for a certain margin of error, these are unquestionably shocking statistics which demand to be addressed. How can the world can remain so incredibly divided within the North and between North and South more generally? (UN 2003).

It has been argued by many Third World writers, and progressive writers in the North such as Richard Falk, that the global political economy remains premised on a form of global apartheid. This presents a very different sense of globalization (with associated characteristics such as global homogeneity) because it is based on an assumption of fundamental inequality and difference rather than uniformity and mutual benefit. It also has implications for how we might understand geopolitics as a theory and practice.

The origin of the term 'apartheid' refers to a policy or system of segregation or discrimination on grounds of race and was introduced by the white minority regime in South Africa in 1948 (see In focus 3.1).

In focus 3.1: Apartheid South Africa

In 1948 the South African government under President D. F. Malan introduced a set of policies and practices which became known in Afrikaans as apartheid (separate development). Over the next 40 years, elaborate plans were constructed not only to identify different racial groups (whites, blacks, coloureds and Indians) but also to develop the South African economy and society along racial and ethnic lines. Politically, white South African citizens were the only category of people able to vote and participate in government. In terms of education, housing, social services and transport strict segregation was enforced. Marriage between 'black' and 'white' South Africans was forbidden and residential areas were demarcated by racial classification. This system of apartheid was condemned by many countries in the United Nations because it actively, and often violently, suppressed the basic human rights of black and other non-white peoples.

In 1990, the most famous political prisoner in the world, the black lawyer (and condemned terrorist) Nelson Mandela was released from detention in South Africa. Over the following years, the white minority government was forced to bow to domestic and international pressure to release hundreds of political prisoners, thereby beginning the process of dismantling apartheid as a prelude to constitutional change. In 1994, the first free and non-racial elections were held in South Africa with the result that Nelson Mandela became the first black president of the country. However, in spite of the formal ending of apartheid, profound inequalities remain between white and black South Africans. Under the leadership of Archbishop Desmond Tutu, the Truth and Reconciliation Commission (TRC), created in the aftermath of the 1994 elections, has attempted to expose the violent nature of apartheid to wider critical scrutiny. In 1999, Thabo Mbeki was elected as the second post-apartheid president of South Africa and Mandela remains a global icon of hope and humility.

We live in a world, as the American strategic thinker Thomas Schelling once noted, where one fifth of the world is rich and predominantly lighter-skinned and four-fifths are poor and darker-skinned. The richer peoples also enjoy an overwhelming military superiority and often seek to prevent the poorer folk (often formally colonized in the past) from 'penetrating' and/or 'swamping' their developed regions (see Schelling 1992). Military force combined with surveillance technologies continues to be used in order to prevent movement of 'economic refugees' and/or 'illegal migrants' from regions such as Latin America and North Africa to North America and Western Europe respectively. Unsurprisingly, various international commissions and reports such as the

UN-appointed South Commission have concluded that the unequal character of the global political economy had to be acknowledged and tackled:

> While most people of the North are affluent, most of the people in the South are poor; while the economies of the North are generally strong and resilient, those of the South are mostly weak and defenceless; while the countries in the North are, by and large, in control of their destinies, those of the South are very vulnerable to external factors, lacking in functional sovereignty . . . And the position is worsening, not improving (South Commission 1990: 1–2).

Absolute poverty and lack of educational opportunities, especially for women and girls, have combined to ensure that millions of people in East Asian countries such as Cambodia, Laos, Mongolia and China have to survive on less than one US dollar a day. Rural and agricultural communities in the East and South Asian region were perceived to be particularly vulnerable to abject poverty. India, for example, has at least 350 million people living in extreme poverty. Non-governmental organizations have, however, often been critical of Northern-dominated international institutions such as the **World Bank (WB)**, the **World Trade Organization (WTO)** and the **International Monetary Fund** (IMF) because of their failure to address village-scale development and urban slum regeneration (Desai and Imrie 1998). In contrast, large-scale projects such as dam construction have tended to dominate the funding profile of international financial agencies for the last 50 years. In the 1990s and beyond, World Bank and United Nations Development Programme reports on poverty and underdevelopment have tended to emphasize the significance of indigenous education spending, gender and infrastructure-led investment without ever considering how North–South relations might impinge upon the capacity of the South to invest in these particular sectors. Moreover, the continued presence of trade barriers and subsidy regimes in the North (such as the Common Agricultural Policy within the EU) perpetuates profound inequalities as the Global South is instructed by the IMF and WTO to 'open up' its economies to international flows of capital.

This chapter is founded upon a belief that Northern debates on global geopolitics (especially with the current concern for 'global terror') and the unequal impact of globalization have either neglected or marginalized the experiences of the South and now former members of the Soviet Union. The future of regions such as Africa, Asia and the Pacific in any new world order will depend upon the interaction of states co-existing within a globalized system of financial flows, social actors, militarization, markets, international organizations and unwanted ideas and threats. The position of countries in Sub-Saharan Africa such as Malawi and Uganda is all the more precarious as it becomes evident that not even so-called Great Powers such as the USA can shape the international system to suit exclusively American needs. This discussion of the South during the post-Cold War era concludes that the North–South cleavage can only be tackled by the progressive strengthening of a global civil society bolstered by an agenda of demilitarization (see Chapter 5), cultural security, sustainable development, environmental protection (see Chapter 6), human rights (see Chapter 7) and global governance (see Walker 1988, Falk 1995).

The 'Third World' and the Cold War

The invention of the 'Third World' by Western social scientists in the early 1950s coincided with the geographical extension of the systemic-ideological struggle between the two **superpowers**. It was perhaps no coincidence that new categories such as 'First World' and 'Third World' were being deployed at a time when the United States and the Soviet Union were directly involved in supporting opposing sides in the Korean peninsula and at a moment when the USA was overthrowing the elected government of Mossadegh in Iran in 1953. Subsequent events in Korea, Vietnam and Central America were increasingly evaluated and judged within a narrative which stressed the significance of the ideological struggle between the superpowers. The geopolitical imagination of the Cold War was characterized by:

> Geopolitical space [being] conceptualised as a three-fold partition of the world that
> relied upon the old distinction between traditional and modern and a new one between
> ideological and free. Actual places became meaningful as they were slotted into these
> geopolitical categories, regardless of their particular qualities (Agnew 1998: 111–12).

In the United States, successive administrations from Truman to Reagan adopted the geopolitical view that the 'Third World' had to be saved from the enduring evils of communism and totalitarianism. In some cases, this concern resulted in armed intervention in various parts of the world, ranging from the widespread carpet bombing of Cambodia in the 1970s to the dispatch of 20,000 marines to the Dominican Republic in 1965. Moreover, other countries such as Israel, Egypt, Taiwan and South Korea received extensive financial and military assistance from the 1950s onwards because the Soviet Union was considered to pose a threat. Taiwan, for example, derived 5–10 per cent of its national income from American financial aid in the 1950s (Ward 1997).

However, American commitments to the Third World were not geographically uniform. Throughout the Cold War, the Middle East, Latin America and the Caribbean were considered to be highly significant whilst other regions such as West Africa were considered to be of lower geopolitical importance (see In focus 3.2).

This geographical variability has been noted in an analysis of the presidential State of the Union addresses between the 1940s and the 1980s (O'Loughlin and Grant 1990, cited in Agnew 1998: 116 and Fig. 3.1). In the early stages of the Cold War, presidents tended to stress the threat to the so-called rimland states which surrounded the Soviet Union and China. In the 1960s, attention tended to be focused on the two socialist states of Cuba and Vietnam. By the 1980s, however, Presidents Carter and Reagan were expressing concern for the Middle East, Southern Africa and Central America.

While the overall pattern of concern may not be surprising given the geopolitical contours of the Cold War, this analysis includes the consistently high priority given to Latin America and the Caribbean by American administrations. This concern for a neighbouring region was rarely benign, however. From 1945 onwards, American administrations developed a range of policies and strategies designed to protect Latin America from socialism and to promote American commercial and security interests. These included the creation of an inter-American security community (under the 1947 Rio Pact), which

In focus 3.2: US support for Israel and the Israeli-Palestine dispute

One of the most controversial elements of US geopolitical strategy during the Cold War was the financial and military support offered to Israel after its formation in 1948. Following the 1917 Balfour Declaration, which declared that a Jewish homeland should materialize, the British as the imperial power were forced to leave the region in the mid-1940s. Jewish terror gangs such as the Stern Gang were highly effective in securing the ousting of British forces.

After the 1948 Independence War, which witnessed the mass expulsion of Palestine Arabs, Israel consolidated its territorial presence armed with the Zionist slogan 'A land with no people for people without land'. In 1967 following a war with Arab neighbours, Israel occupied the Sinai Peninsula in Egypt and the Golan Heights in Syria. In 1982 it invaded and occupied South Lebanon. According to supporters of Israel, the USA (and France) was right to help Israel maintain its political existence given the experiences of the Jewish Holocaust and persistent hostility from surrounding Arab states. Israel remains an undeclared nuclear power and unlike its Arab neighbours, a parliamentary democracy.

For the critics of Israel and its support from the USA, this policy has allowed the country to ignore UN Resolution 242 (1967), which calls for a 'just settlement'. Seven hundred thousand Palestinian Arabs were exiled into Jordan and millions more live in miserable conditions in the West Bank and Gaza Strip. Palestinian terror groups targeted Western and Israeli individuals and the state apparatus as part of their campaign for international recognition. The Palestinian leadership continues to push for a full and final territorial settlement with Israel.

As part of the gradual improvement in relations between Israel and the Arab world, Egypt recognized Israel's right to exist in 1982 and in return Israel left the Sinai Peninsula. The Oslo Peace Process (1993) and subsequent negotiations such as at Wye (1998) have been plagued by terrible violence as Israel seeks to consolidate its grip on the West Bank and the Gaza Strip in response to Palestinian resistance known as the *Intifada*. Suicide bombers have targeted Israelis (often on commuter buses) and thus many Israelis support a repressive policy against the Palestinians. It is hoped that in 2005/6 an independent Palestine will exist in return for guarantees regarding Israel's right to exist in the region. The prospects remain bleak, not least because it remains unclear whether both sides can agree on territorial boundaries, the control of Jerusalem, the right of return for Palestinian exiles, and the fate of illegal Israeli settlements in the West Bank and Gaza.

involved mutual defence in the Americas and the provision of financial and military assistance through programmes such as Alliance in Progress in the 1960s.

In more extreme cases, however, the American military and intelligence agencies were prepared to undermine governments in the Latin American region considered to be leaning towards the political left. In 1954, for example, the Central Intelligence Agency (CIA) provided rebels in Guatemala with funds, arms and combat training so that they

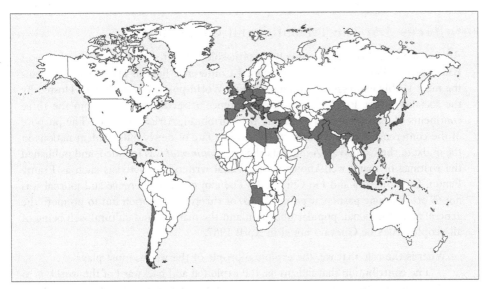

Figure 3.1 US–Soviet conflict: zones of most serious trouble, 1948–88
Source: Adapted from Nijman 1992

could successfully overthrow the reformist government of Jacobo Arbenz Guzman (Immerman 1982). In 1961 the CIA also encouraged rebels to attempt an overthrow of the socialist regime of Fidel Castro. The Kennedy administration of the time provided arms to Cuban rebels and promised US air support to encourage a coup against President Castro. In April 1961 a rebel force landed at the Bay of Pigs only to find that Castro's military forces hopelessly outnumbered them. US air support never materialized and the subsequent failure of the so-called 'Bay of Pigs' venture was not only a crushing revelation of the limitations of American power but also contributed to the worsening relations between the superpowers over Cuba. The decision by the Soviet Union to place missile installations on Cuba precipitated one of the tensest moments of the Cold War when it appeared that the United States was prepared to launch military strikes against Cuba if the installation work continued. The crisis eventually ended when Soviet missile transporters were returned to their home bases and the Americans agreed to withdraw their Jupiter missiles from Turkey.

In the same year as the Bay of Pigs fiasco, Third World states came together as a political force. The creation of the Non-Aligned Movement (NAM) in 1961 was an illustration of how some Third World states attempted to resist the international politics of the Cold War (Willetts 1978 and In focus 3.3).

Composed of states such as India, Egypt and Yugoslavia, it was hoped that the NAM would contest the geopolitical pressures of the superpowers. Non-alignment is not the same as neutrality because the latter is usually a condition which is recognized or guaranteed by other states. Non-alignment is concerned with developing an independent political space which is secure from superpower interference. The founders of the

---

In focus 3.3: The Tricontinental Conference

After a successful conference involving post-colonial states in the Indonesian city of Bandung in 1955, the 1966 Tricontinental Conference held in Havana was perhaps the most visible expression of militant Third World politics in the 1960s. Hosted by the socialist leader Fidel Castro, the conference attracted delegates from the three continents of Latin America including the Caribbean, Africa and Asia. The purpose of the conference was to consider the collective fate of newly independent nations in the midst of the Cold War. A journal called *Tricontinental* was founded, and published the writings of many well-known post-colonial writers and activists such as Frantz Fanon, Che Guevara and Ho Chi Minh. The aim of the conference and journal was not to produce one particular political and/or theoretical position but to promote the general aim of national/popular liberation and the material and cultural well being of all peoples. As Che Guevara noted in April 1967:

What is the role that we, the exploited people of the world, must play?

The contribution that falls to us, the exploited and backward of the world, is to eliminate the foundations sustaining imperialism: our oppressed nations, from which capital, raw materials and cheap labor (both workers and technicians) are extracted . . . sinking us into absolute dependence. The fundamental element of that strategic objective, then, will be the real liberation of the peoples.

Descriptions such as 'tricontinental' are not just geographical labels but they also serve to remind us that there are alternative viewpoints and knowledge systems about the state of the world. *Tricontinental* sought to change the values and terms under which we live.

Source: Young 2003: 17–18

NAM in 1961 tried to create a political forum in which common problems such as the building of a new state in the midst of the Cold War could be discussed. Over the years, the NAM met at intervals of over three to five years to consider the political and economic issues: Cairo 1964, Lusaka 1970, Algiers 1973, Colombo 1976, Havana 1979, New Delhi 1983, Harare 1986, Belgrade 1989, Bogota 1994, for example. South Africa joined the NAM in the same year that President Mandela was elected the country's first black president in 1994. Although the NAM had no central headquarters, the group did co-ordinate activities on technical co-operation, development, disarmament and international security. Summit meetings were the major venues for debate and policy formulation (Singham and Hune 1986).

At the 1973 NAM Summit, the parties committed themselves to pursuing a New International Economic Order (NIEO) in order to reduce the North–South divide. This NAM Summit in conjunction with the raising of oil prices by OPEC in 1973 prompted discussion of the NIEO at the UN in 1974. Despite the high profile of the NIEO debates, the NAM never really enjoyed high-level political success because its members were divided on the ultimate objectives of non-alignment. Some countries such as Cuba and Libya wanted the NAM to align itself more closely with the Eastern Bloc, whilst others argued that the movement should look to the West for political support. By the late 1970s, arguments for a NIEO had declined in political salience not least because the re-emergence of a Second Cold War had shifted the political agenda away from economic issues. With the ending of the Cold War in the late 1980s, the political significance of NAM largely disappeared. The organization continues to meet in order to discuss the politics of non-alignment (arguably in a context shaped by the US-led 'war on terror') in the early part of the twenty-first century.

These struggles for survival should not be underestimated given the scale and intensity of violence in many parts of the Third World (see In focus 3.4).

In South East Asia, for instance, over 600,000 local people died due to confrontations between rival American and Soviet-backed military forces between 1969 and 1975. Intelligence agencies such as the CIA also pursued an assassination programme,

---

In focus 3.4: Chile's September 11th

The United States is not the only country to have suffered extreme cultural trauma associated with 'September 11th'. Chile's particular trauma occurred with the overthrow of the elected socialist government of Salvador Allende by a military coup on 11 September 1973. Supported by the CIA and American corporations such as ITT, the military *golpe de estado* was prompted by a fear amongst the Chilean military and business sectors that Allende's socialist social and economic programmes would either ruin Chile's economy and/or ensure that Chile became a client state of the Soviet Union. The Americans, already fearful of Castro's Cuba in the Caribbean, were determined that socialism should not gain a foothold in Latin America. As former secretary of state Henry Kissinger once remarked, a country like Chile would not be allowed to 'go Marxist' just because 'its people were irresponsible' (Hitchens 2001: 55). On 11 September 1973, the presidential Palace in Santiago was stormed by Chilean armed forces under the leadership of General Augusto Pinochet and Allende was later killed. Pinochet assumed the political leadership of the country and governed Chile for the next seventeen years. As part of his determination to prevent any future 'political Allendes' he pursued a vicious campaign against any individual or group suspected of having left-wing leanings. It is estimated that at least 3,000 people were murdered by the military regime and in neighbouring Argentina the death toll was even higher as a military regime pursued its own 'war of terror' in the late 1970s.

Ironically, in December 1998 former president Pinochet was arrested in London for the purpose of facing charges from a Spanish court relating to mass murder and human rights abuses. He was eventually released by Britain on the grounds of severe ill health. Notwithstanding his escape from international justice, the episode did demonstrate that former heads of state no longer enjoy automatic immunity from prosecution.

Source: Hitchens 2001

'Operation Phoenix', against Vietcong supporters in the early 1970s. In other parts of the world, socialist and military regimes in Africa, Latin America and Asia strove to consolidate the powers of the state within a rapidly changing world economy. Socialist governments such as Mozambique and Angola were racked by civil wars and superpower intervention (in Southern Africa) in the 1970s. Over 1 million people are believed to have died between 1975 and the early 1990s in Mozambique alone (Sidaway and Simon 1993, Power 2003). International agencies such as the World Bank had to provide emergency financial aid in order to save these states from total collapse due to civil war which also destroyed the early achievements in health care and education provision.

The NAM succeeded in changing the often violent profile of North–South relations through its adoption of a campaign for a NIEO based on financial and technological transfers from North to South and through the promotion of peaceful co-operation between states (Thomas 1987, Halliday 1989). The initial impetus for a NIEO stemmed from the development at the United Nations Conference on Trade and Development (UNCTAD) and the creation of the Group of 77 within the United Nations in 1964. The Group of 77 represented the poorest member states of the UN and was designed to bring Southern voting power to bear on Northern member states of the UN Security Council. The meetings of the UN General Assembly and the UNCTAD were used to raise the issue of unequal trading relations between North and South. Demands for a NIEO were based on a belief that radical change was needed in order to improve the condition of the South. Basic demands included: a new general system of preferences to enable the South to break into the manufacturing markets dominated by the North; a commitment from the North to devote at least 1 per cent of GDP to official aid; the cancellation of the 'Southern' debt; technology transfers to be executed; and the improvement of control and regulation of multinationals to prevent the exploitation of Southern resources and labour markets.

This was an ambitious agenda, which demanded radical reforms of the international economic order. It was also conservative in the sense that co-operation between states was still considered to be the best means of promoting economic development for the South within the capitalist world economy. However, it was also grounded on a belief that structural obstacles within the global political economy would have to be overturned. In the late 1970s, there appeared to be some evidence that the South was making progress and that even the UN-appointed Brandt Commission (named after the former German Chancellor Willy Brandt) recognized the significance of these inequalities between North and South. Furthermore, the South proved to be an effective negotiating bloc during the oil price rises crisis of 1973–4 and the United Nations Convention on the Law of the Sea in the 1970s and 1980s. The declaration of the ocean floors as common heritage (and therefore the property of the global community) was a considerable political success despite American and Northern opposition. However, fundamental change in the world economy was elusive in the 1980s as priorities changed and the onset of the Second Cold War ensured that Northern states were more concerned with rising superpower tension than North–South relations. By the time of the 1982 World Summit of Northern and Southern leaders in Mexico, it was abundantly clear that Northern leaders such as President Reagan and Prime Minister Thatcher had no interest in meeting the demands of the NIEO.

The Northern states' apparent lack of interest in fundamental reform led Southern states and their commentators to talk of a so-called 'lost decade of development' (see Green 1995). Throughout the 1980s, the political and economic condition of many parts of the Third World began to worsen as economies collapsed in Sub-Saharan Africa and Central America witnessed the long-term destabilization of Nicaragua and the 1989 invasion of Panama. The renewed geopolitical confrontation between the Soviet Union and the USA had, therefore, dire consequences for the economic and political welfare of the Third World. Armed intervention in combination with rising debt burdens and public-service sector collapse prompted discussions of so-called 'failed states', a term first introduced in the 1980s to convey a sense of places where the basic mechanisms of governance had simply evaporated. For Mozambique, governance was increasingly determined by international bodies based in Washington DC rather than in the national capital of Maputo (see In focus 3.5).

In focus 3.5: Political conditionalities and the 'Washington Consensus'

In 1991 the United States, Britain and multilateral donors introduced so-called 'political conditionalities' for the purpose of securing 'good governance'. These demands were labelled the 'Washington Consensus' because they originated in the United States and US-based international institutions such as the IMF. In order to qualify for loans, countries had to, amongst other things, curb budget deficits, reduce public spending, protect property rights, liberalize trade, privatize state-owned corporations and promote foreign direct investment. The stipulations regarding 'good governance' were defined by the donors, and thus in conjunction with the 'economic conditionalities' attached to structural adjustment programmes (SAPs), this could be seen as yet another attempt to undermine the sovereign authority of impoverished states in the Global South.

By the end of the Cold War, the NAM had lost its economic and political appeal because of the changing relationships between its members, the superpowers and the wider international community. The onset of the debt crisis in 1982 (see below) further compounded the South's inability to demand fundamental change in spite of the initial shock to the Northern financial community. Within the Southern coalition, collective demands for radical reform were also beginning to fragment as it became apparent that some states such as South Korea and Malaysia had enjoyed considerable success in terms of economic growth and rates of industralization. For world-systems theorists, the growth of a Southern semi-periphery was a natural outcome in the sense that the world economy needed economic and political safety valves. It was therefore in the North's interests that some Southern countries developed successfully whilst others remained underdeveloped. The rapid political changes of the 1980s induced some analysts and political leaders to argue that the South or the 'Third World' had effectively ended because of the diversity of experience in the regions. New times demanded new political programmes and new forms of analysis.

The end of the Third World?

Since the end of the Cold War and the collapse of the Soviet Union, increased attention has been paid to the intellectual utility of Cold War categories such as First and Third Worlds. It has been widely suggested that the term 'Third World' is no longer an appropriate label for the complex and varied regions of North Africa, South Asia, Sub-Saharan Africa, Latin America and the Caribbean, South East Asia, South West Asia and the Pacific (Berger 1994, Ayoob 1995, Grant 1995, Haynes 1996). During the 1990s, critical observers in the North and South advanced three major objections to the concept of a Third World (Fig. 3.2).

The first could be described as a philosophical objection to the implicit assumption of three different worlds (Hosle 1992). The concept of a Third World erroneously implied that the lives of human beings in Africa, Asia and Latin America were entirely separate from those living in the First and Second worlds. As globalization theorists have stressed, all human beings live in one and only one interdependent world. The formation of an industrialized North and an underdeveloped South was intimately related rather than derived from separate economic and political processes. Moreover, the differentiation

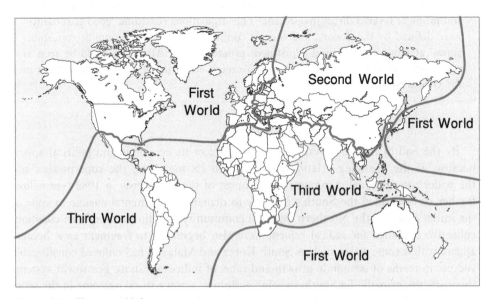

Figure 3.2 Three worlds?

between First/Second/Third worlds implicitly assumed a value hierarchy where the first is considered superior to the third.

During the Cold War, the term 'Third World' had an apparent analytical value because it seemed to refer to states which not only shared a common colonial experience but were also intent on economic development. Mainstream development approaches in the United States ensured that such a categorization also implied that the Third World should be seeking to follow the example of the First World. Walter Rostow's manifesto for a non-communist approach to economic development, for example, assumed that there were five major stages of development, which would involve a substantial transformation in the cultural, economic and political life of developing nations (see Desai and Potter 2002). In the process, it was generally assumed that development would be a relatively uniform process for the Third World states regardless of their particular location and history. The division of the world into three separate spheres meant in practice that Western observers tended to neglect the interrelationships between these allegedly separate worlds.

The second point of objection is concerned with the ending of the Cold War. The concept of the Third World was developed in the 1950s by Northern social scientists to refer to a world dominated by the bloc politics of the Cold War. A tripartite division of the world made some sense in the 1960s when the world was characterized by a superpower confrontation and the emergence of newly independent nations in Africa and Asia. However, these circumstances changed radically and in alliance with the acceleration of political and economic globalization, the world has witnessed the rapid transformation of the earth's political geography. Some parts of the Third World have become highly developed while others have floundered. Until 1997–8, the so-called East Asian tigers of Malaysia, Singapore and Thailand experienced some of the highest economic growth rates in the last twenty years (see In focus 3.6).

The collapse of 'Second World' federations such as the Soviet Union and Yugoslavia has meant that some of the former Soviet republics such as Armenia and Azerbaijan are alleged to resemble Third World economies (see Bradshaw and Stenning 2004). More generally, a shift of geo-economic influence from the Euro-American realm towards the Asia-Pacific basin has meant that the political geography of the post-Cold War era is quite different from that of the 1950s and 1960s.

One major element of change in the political geography of the world economy has been the rising profile of China, which has been described as the next economic and political superpower after the USA and Japan. As early as 1975, the *Economist* magazine was predicting that China's expanding economy would be a major force in the world economy. To date, China's economy has grown at around 10 per cent per annum since 1991 and it now produces half the world's toys, two-thirds of its shoes and most of its bicycles and power tools. China is also the largest recipient of foreign investment after the USA. Economists estimate that China's GNP (Gross National Product) could exceed that of the United States (the largest economy at present) by the end of the twenty-first century. Since the 1990s, China has engaged in a massive programme of market reform and commercial development. There is little doubt that standards of living have improved for many Chinese people in terms of access to clean water, possession of

---

In focus 3.6: The 1997–8 Asian tiger crisis

In July 1997 the Thai currency, the baht, collapsed in the midst of a general economic downturn involving the largest Asian economy, Japan. Annual GDP growth for the South East and East Asian region in 1997–8 went into rapid decline as a loss of investor confidence meant that $100 billion was withdrawn from South East Asian economies. Within a matter of months, Indonesia, Malaysia and the Philippines suffered further reductions in economic growth as currencies collapsed and international financial confidence evaporated. Indonesia, despite an abundance of population and resources, had a debt of $180 billion at the same time as it underwent a massive political upheaval with the overthrow of the authoritarian government of President Suharto. In Malaysia, the Mahathir government imposed capital controls during 1997–8 in an attempt to prevent international capital flight. In the Philippines, a weak government led by Joseph Estrada failed to prevent a large-scale reduction in GDP and the national currency, the peso, plunged to an all-time low while poverty levels increased markedly. Foreign exchange reserves of countries such as the Philippines were used in an attempt to halt the collapse in domestic currencies. Singapore was able to respond more effectively to the economic crisis than the Philippines because a stable national government was able to halt the substantial decline in the Singapore dollar and contraction of its export sector by raising interest rates and bolstering investor confidence. The IMF had to provide international financial credit for the region in order to assist its general economic recovery, although there remains considerable controversy over the wisdom of this intervention in the light of earlier IMF advice to 'open up' these economies to foreign direct investment.

---

consumer goods and better food and housing. However, the environmental and social costs have been high in terms of poor employment conditions for many workers, water shortages, environmental degradation due to industrial pollution and continued controversies over the state of human rights in the country including the disputed region of Tibet.

The final point of objection to the term 'Third World' concerns the elites within these states. The promotion of Third Worldism in the 1970s and 1980s disguised the fact that Third World elites (often Western-educated) were not always acting in the best interests of their own societies. Notorious political leaders such as former President Idi Amin of Uganda (trained at Britain's elite military college, Sandhurst) stole millions of pounds and dollars from their governments and deposited the money in secret Swiss bank accounts. In the early 1970s Amin attempted to either kill or expel all the ethnic Asian Ugandans in a bid to ethnically cleanse Uganda of 'foreign' elements. In spite of its rich natural resources and exports such as coffee, Uganda is now one of the most heavily indebted countries in the world relative to the size of its economy. In the Central African Republic, the former self-styled 'Emperor' Bokassa spent $20 million (equivalent to 25 per cent of the total GDP) on his coronation ceremony in 1977. In Zaire, former President Mobutu stole several billion US dollars over a period of 20 years, which was derived from the country's export in oil and diamonds (Reyntiens 1995). Categories such as 'Third World' effectively homogenized conditions within these parts of the world rather than exposing the enduring and contradictory complexities of these post-colonial societies. Within the socialist world of Third World states, high levels of violence directed against an internal population often overwhelmed appeals to equality and socialist forms of development. The 'killing fields' of Kampuchea (now Cambodia) in the 1970s are a chilling reminder of how a socialist regime led by the Khmer Rouge leader Pol Pot participated in the massacre of 2 million people.

Far from ushering in a new global order based on uniform economic development and liberal democracy, the conditions of the Third World remain so varied that the standard social science categories such as 'developing countries' and the 'periphery' increasingly do not make sense for countries ranging from Cambodia to Yemen and from Singapore to Togo. Robert Gilpin noted in 1987 that the Third World 'no longer exists as a meaningful entity' (Gilpin 1987: 304). Rapid political change has, therefore, apparently called into question the capacity of mainstream concepts and theories to explain and interpret the world around us. As Cedric Grant has claimed:

> Since the collapse of communism and the end of the Cold War, the scepticism as to the existence of the Third World has increased. This is because the term 'Third World' was derived in the context of a bipolar world as a label to differentiate the newly independent countries of Africa and Asia from the rival power blocs, the Western and the Soviet, which in their competition with each other had focused their attention on these newly independent nations. Even those who were inclined to agree that there was some substance to the concept of the Third World are now more ready to accept the contention that the global transformation which is occuring is rendering the concept anachronistic (Grant 1995: 567–8).

The actual delimitation of a 'Third World' during the Cold War deserves further elaboration because it touches upon some of the enduring controversies surrounding those countries in Africa, Asia and Latin America which have yet to achieve economic wealth comparable to that of Western European and North American states and Japan. In addressing these questions, this discussion will broach the intellectual and academic context which gave rise to the concept of a Third World. The purpose of this investigation is to demonstrate that the conceptual challenges posed by the concept of a Third World are far greater than a simple presentation of the changing political map in the post-Cold War era. In an era of increasing globalization, the advocacy of a concept such as the Third World could be used to promote a spurious impression of homogeneity, thereby reproducing an unhelpful distinction between a First and a Third World. On the other hand, the term 'Third World' can be useful in highlighting the persistent inequalities within the world system and the enduring aspirations of several billion people.

With any label such as 'Third World' or 'Developing World' or 'Low Income World' there are always inherent difficulties in representing either vast areas of the earth's surface or complex socio-economic situations in terms of single categories (Barton 1997: 6).

The term 'Global South' is preferred because it is a geographical reference to the southern hemisphere, which in spite of the inclusion of countries such as Australia, South Africa and New Zealand, is overwhelmingly the poorer hemispheric region of the world. The Brandt Commission acknowledged this feature in the early 1980s when it identified a North–South divide in reports on world development. The term 'Global South', therefore, is intended to highlight similar economic, environmental, social and political conditions whilst recognizing that Southern regions are complex and diverse (and that the populations of India and China live north of the equator!).

US–Latin American relations, debt burdens and the ending of the Cold War

The British geographer Doreen Massey employed the term 'power-geometry' to highlight the unequal and paradoxical nature of globalization (Massey 1991). On the one hand, Northern governments and financial commentators frequently depict the earth as a world of unfettered spaces, whilst on the other hand they also seek to control and regulate movement and flows within bounded spaces (see In focus 3.7).

The immigration controversies in the USA and Western Europe reveal the desire of rich countries to restrain the movement of poorer peoples while simultaneously demanding the free movement of capital and investment. In California in the 1990s, for instance, voters were debating Proposition 187, which set out to prevent illegal immigrants from accessing any form of public service such as health and welfare. Yet at the same time, these illegal immigrants provide services such as office cleaning and strawberry picking which the local populace was unwilling to perform because of poor pay and, in the case of soft-fruit harvesting, the 'back-breaking' nature of the work. These spatial inequalities ensure that the poorer regions of the world are held in place and invaded by the rich in terms of economic investment and political interference. For poorer regions such as Latin America, the ending of the Cold War has not radically changed the political-economic condition of the population. As a Mexican political

---

In focus 3.7: Two types of boundaries?

Advocates of hyper-globalization espouse a 'borderless' model of the world in which national borders should not impede the free movement of capital, trade and ideas. Northern states frequently condemn others, especially in the Global South, for preventing the free flow of trade and capital.

Alternatively, when it comes to the free movement of people, borders often take on a renewed significance. Proponents of 'Fortress Europe', for example, contend that people (often defined as illegal immigrants or economic migrants) cannot be allowed to cross borders in order to search for work. It appears acceptable for capital and trade to flow freely but not for people to move across borders.

scientist has noted: 'Latin America . . . finds itself in a sadly paradoxical bind. The end of the Cold War has brought greatly broadened geopolitical leeway, but economic globalization and ideological uniformity have rendered that at least partially meaningless' (Castaneda 1994: 48).

Investigating the role of the South in the post-Cold War era is a necessary component of any critical evaluation of globalization. The South, as Jonathan Barton has argued, cannot be considered to be peripheral to such an investigation (see Barton 1997). In Nicaragua, a country caught up in the ideological and territorial struggles of the Cold War, per capita income has fallen in real terms as a result of economic pressure from the North, geopolitical destabilization by rebel forces and US military support of anti-government forces. In Guatemala, where 2 per cent of the population own 60–70 per cent of the most productive land, the ending of the Cold War did not lead to a transformation of land ownership. Moreover, the US invasion of Panama in 1989 reminded Central Americans that the sole remaining superpower has never been averse to violent intervention in the region when it wished to re-secure regional hegemony. The removal, with the help of 10,000 US troops and loud rock music (used to 'bombard' the presidential palace), of the country's leader General Noriega (whom the US accused of condoning an extensive drugs trade), was ironic given the US's previous support for the military leader. Other commentators have also pointed to the fact that the US was concerned about growing levels of Japanese investment in the Panamanian Isthmus and was thus anxious to restore its geopolitical authority over the area. The Panama invasion was a significant development as it was the first hostile post-Cold War incursion. As the Honduran newspaper *La Tiempo* noted in December 1989:

> It was a coarse grotesque euphemism [Operation Just Cause: the code name for the
> American invasion], neither more nor less than an imperialist invasion of Panama . . .
> We live in a climate of aggression and disrespect . . . hurt by our poverty, our weakness,
> our naked dependence, the absolute submission of our feeble nations to the service of
> an implacable superpower. Latin America is in pain (cited in Chomsky 1991: 158).

The invasion of Panama coupled with massive destabilization of Central American governments by the superpowers contributed to the so-called 'lost decade' of development and social progress in the 1980s.

The failure to eradicate the debt burden of the Global South is probably the single most enduring inequality between North and South. In 1990 it was estimated that the total debt of the South/Third World had reached $1.5 trillion dollars. In Latin America, the debt burden accounted for a substantial amount relative to total export earnings: Mexico $85 billion, Brazil $105 billion and Argentina $61 billion (1998 figures). The most indebted continental region remains Sub-Saharan Africa when measured by total external debt in relation to the export of goods and services (Simon *et al.* 1995). Through a combination of factors including the rapid rise in lending by Northern banks and states in the late 1970s, Southern states accumulated substantial debts by the 1980s because of their incapacity to repay loans and grants. Global economic depression in the 1980s further contributed to this so-called lost decade of development for Latin

America and Sub-Saharan Africa. The suspension of debt repayment by Mexico in August 1982 precipitated the biggest financial crisis in the history of the international financial system. Shortly afterwards, other states such as Brazil and Argentina suspended their debt-repayment schedules too.

Over a period of 15 years, the international community has promoted a range of debt-rescheduling packages for countries such as Mexico. With the assistance of the US, the Mexican government was instructed by the World Bank to follow an austerity package which sought to devalue the national currency and cut public spending in order to reduce the annual burdens on the Mexican treasury. However, after a decade of austerity the country was hit by further financial crises which led to the collapse of the peso, the withdrawal of foreign investment and a decline in economic growth. In 1998 the Mexican debt was estimated to be $85 billion, at a time when a new debt-relief deal with the World Bank and IMF had been envisaged.

The recent experiences of Mexico have been repeated, admittedly in different ways, around the countries of the South. Attempts to structurally adjust debt-ridden economies have not been successful in promoting sustainable development or reducing poverty and hunger in the South. The idea of structural adjustment policies was to liberate extra monies for debt repayment through public-sector reductions in spending. This has not been effective in terms of building a more sustainable future for Southern societies because economic plans tended to emphasize reductions in consumption rather than investment for people in the future. In Latin America, the US has been actively involved in reducing debt levels (in a somewhat piecemeal fashion) because of the geographical and political-economic proximity of countries such as Mexico. It has been argued, for instance, that American plans to create a North American Free Trade Association (NAFTA) depended, amongst other things, on Mexico's financial position being improved by the 1980s. Debt-relief plans for Mexico were implemented by the Reagan administration to increase confidence in the Mexican economy. President Carlos Salinas de Gortari of Mexico later claimed that an 'economic miracle' had occurred between 1988 and 1994 because of the rise in foreign investment in the form of speculative capital.

The subsequent financial crisis in Mexico in the mid-1990s sparked off a wave of protests against structural adjustment and debt burden. In January 1994 a guerrilla uprising by the Zapatista National Liberation Army (EZLN) in the southern state of Chiapas coincided with Mexico's formal entry into the North American Free Trade Association (see Chapter 8 for more details). Such expressions of dissent and resistance were perhaps unsurprising given the accumulating evidence that **structural adjustment programmes** (SAPs) and free-market reform packages were worsening the social and economic condition of the poor, rural inhabitants and the plight of women and children. Levels of inequality and opportunity have worsened in reformed economies such as Mexico. The current president, Vicente Fox, continues to press ahead with 'reforming' the Mexican economy (as part of international and regional pressures from the IMF and the US and NAFTA respectively), assisted by political support from the United States despite worsening social polarization (see In focus 3.8).

In focus 3.8: The collapse of Argentina?

In Argentina, the application of a SAP in combination with a high debt burden led to widespread rioting and political meltdown in 2001–2. Ironically, former Argentine president, Carlos Menem, had been acclaimed for the successful economic transformation in the 1990s when the local currency (the peso) was pegged to the US dollar, inflation was controlled and a widespread privatization programme was initiated. Within two years of his leaving political office, Argentina was plunged into turmoil as foreign debt reached $130 billion and unemployment was over 20 per cent. The IMF, backed by the US government, demanded that the Argentine government radically reduce public expenditure and ensure that spending actually matched revenue collection in the form of public taxation. With cuts in expenditure on education, health, unemployment benefit and social security, millions of Argentine citizens found that essential social services were reduced and many public-sector workers were simply not paid.

All sectors of Argentine society were affected and many middle-class public-sector professionals such as doctors and academics engaged in widespread public protests that led to the collapse of the Argentine political system. Within the space of two months in 2001–2, Argentina had five different presidents. The Italian and Spanish embassies in Buenos Aires were besieged as many middle-class Argentines with second passports sought to escape the economic and political crisis. An extraordinary intellectual and financial flight occurred and the long-term prospects for the country (the sixth richest in the world in 1900) remain bleak.

Southern views on development, world politics and the debt crisis

For the last fifty years, official development policies have tried to promote development through the political and economic transformation of states in the South (see Escobar 1995, Rist 1997). It could be argued that, by any conventional indicator of development, these policies have failed. In 1997 it was recorded that in 19 countries per capita income had fallen below the 1960 figure. Poverty and hunger continue to affect vast areas of the world including ethnic minorities, the disabled and the elderly in the North. Over 1 billion people still do not have access to clean water supplies and it has been estimated that in terms of global income distribution, well over three-quarters of total income is owned or enjoyed by the richest quarter of the global population (UN 2003). In that sense, World Bank figures for GDP (which do not consider patterns of distribution) tell us little about the lives of people living in slums, nor do they remind us that far more people have died from disease and hunger than the 187 million people who perished through wars and conflict in the last century (Hobsbawm 1997).

There is a lengthy if neglected tradition concerned with the actual conditions of the South within the global political economy (Galeano 1973, Love 1980). 'Southern' views of international politics have been constructed on a more general account of the centre–periphery relationship within the world economy. These accounts are 'Southern' in the sense that the writers hail from Latin America, Africa and Asia rather than the Euro-American world. In the 1950s, for example, the economic writer Raul Prebisch, an Argentine economist working at the United Nations Economic Commission for Latin America, proposed that the North and the workings of the capitalist world economy were restraining the industrialization of the South. He argued that the South's dependence on the production of primary products for the North coupled with the consumption of goods manufactured in the North was inherently disadvantageous to the South. In the long term, trading conditions force the South to derive ever more credit from primary exports in order to retain purchasing power. Unlike manufactured goods and services, primary products do not provide much scope for innovation and increased profitability. For many Southern states, therefore, there is little alternative than to retain their economic and political position in a Northern-dominated international economic order.

In the 1960s, new writers such as A. G. Frank and F. Cardoso (a former president of Brazil) directed the focus of analysis towards class relations and patterns of exploitation. One of the key areas of debate was the extent to which Southern capitalists and governments were junior partners in a global system of exploitation and domination. In his path-breaking analysis *Capitalism and Under-Development in Latin America* (1971), Gunder Frank presented a detailed account of the systematic underdevelopment of the South. In essence, Frank claimed not only that the promise of economic development for the South was inherently false but also that the South was actually participating in its own underdevelopment. The structural constraints on the South were such that economic development was always likely to be minimal and precarious because of the Northern domination of the world economic order. These kinds of ideas, though later criticized for their economic and political assumptions about class, the state and the world economy, were emblematic of a wider concern for the condition of the South. The demands for a NIEO in the 1970s could be attributed to the work of structuralists such as Frank and Cardoso.

Although these accounts of the global political economy have been criticized over the years, the dependency writings contributed to a rather different series of perspectives on international relations. For much of the post-war period, the disciplines of geopolitics and international relations have been resolutely Anglo-American in the sense that most of the Northern-based writers were concerned with either the North and/or the international system per se. Following from this body of literature, **world-systems theorists** such as Immanuel Wallerstein and Peter Taylor argued that social and political relations between the North and South need to be considered within a longer time frame of an evolving capitalist world economy (Wallerstein 1980, Taylor and Flint 2000). The conditions of the Global South in the twenty-first century, therefore, have to be investigated as part of a longer historical process. Governments in postcolonial Africa and Asia have tried to secure their vulnerable national territories and economies in the face of weak state sovereignty. During the Cold War, for example, many nations of the Third World experienced direct interference and military intervention from outside powers seeking to undermine a particular regime. The human cost of these interventions was very high as nations such as Mozambique and Angola were destabilized with dire consequences for civilians, particularly women and children.

Northern debates over globalization have been intensely concerned with the erosion of state sovereignty and transboundary political and economic flows. In the South, experiences of this kind have been routine (since the fifteenth century) in terms of the undermining of state jurisdiction and the penetration of Western influences into national cultures. Mohammed Ayoob and Caroline Thomas have argued that the economic dimensions of national security such as access to secure systems of food, health, money and trade are major concerns for Southern states (Thomas 1987, Ayoob 1995). No wonder then that governments of the South have often been staunch supporters of the principle of non-intervention, mindful of the fact that the international system is not based on the premise of equal and self-determining sovereign states (see Chapter 7). States such as the USA have been far better equipped to deal with the demands of international politics and globalization, whereas others such as Sudan and Mozambique might best be described as quasi-states in the sense that their continued existence and legitimacy have more often than not been derived from international relations rather than internal support (Sidaway 2002). Recent debates over human rights, societal security and humanitarian intervention in the 1990s had substantial implications for the South and its capacity to prevent further erosion of the right of Southern states to conduct their own affairs. Perhaps we should talk of in-dependence rather than independence.

It has became apparent that a number of pressing issues confronting the South and South–North relations have still not been resolved in a satisfactory manner: the political and economic consequences of development, gender and human rights, environmental protection, debt reduction and the protection of ethnic and religious minorities (Haynes 2002). At the same time, mainstream development approaches have failed to tackle the underlying structural causes of poverty, hunger, disease and chronic indebtedness. Major international conferences and meetings such as the 1992 Rio Summit, the 1995 Conference on Socio-Economic Development, the 2002 World Summit in Johannesburg and the 2003 WTO meeting in Cancun have tended to reaffirm a public commitment by the North to the promotion of free trade, market integration and liberal democratic governance, but for 'Southern' critics and NGOs, these forums do not confront the profound inequalities of the global political economic system. The 2003 WTO meeting collapsed because states such as India, China and Brazil complained that the US and Europe were not prepared to end subsidies to domestic farmers. Moreover, Southern critics have expressed anger at Northern critics who blame Southern population increase for global environmental change rather than acknowledging the massive consumption of raw materials by the North.

In contrast, attention in the South has focused on promoting local forms of development which stress local needs, self-reliance, ecological sustainability and community survival. Southern NGOs in alliance with Northern NGOs and progressive commentators have called for new forms of development strategies. Local groups such as the Chipko movement in India and the rubber tappers' movement in Brazil have been lauded for their campaigns to protect access to their environments and resources. Other groups in Guatemala and Ecuador have highlighted the importance of land reform in these countries, where the vast majority have no means of growing their own crops and developing sustainable lifestyles. South Korean farmers have protested against the unregulated flows of American-subsidized rice, which has had a devastating impact on local farming incomes. For the poor of the South, sustainable development is a fiction when rich minorities control most of the fertile agricultural land, leaving the poor in places such as Brazil (where 1 per cent of the population owns 48 per cent of the land) and Zimbabwe (where 2 per cent claim 60 per cent of the land) to exploit fragile uplands and/or rain forests in order to meet their needs.

Conclusions

In the South, the recent transition towards market-based economies and liberal democracies has often been fraught. For one of the poorest countries in the world, Mozambique, the transition from a socialist developmental project to capitalism has been deeply problematic given the state of the country after 20 years of civil war and external intervention. Mozambique's economic and political condition remains parlous even with the ending of the civil war in the early 1990s and recent elections. The destruction of basic education and health provision provides a grim reminder of the profound differences between North and South. Although forms of entrepreneurship and private-sector growth occur in Maputo, the majority of the population remains impoverished and unwanted by South Africa, which constructed an electrified boundary fence in order to prevent illegal migration from the state. 'Fortress South Africa' co-exists uneasily with the apparently unregulated flows of refugees and migrants from southern Africa.

In terms of globalization and geopolitics, this chapter on North–South relations disturbs simplistic assumptions about a world divided (in the form of global apartheid) into an impoverished South and a rich North. The architecture of division is more complex, as some parts of the North are as disadvantaged and socially excluded as the South. While Los Angeles is the second largest 'Mexican' city, the movement of immigrants continues to blur the spatial and imaginative boundaries between the North and South. The mortality rates for Afro-American children in the United States are as horrendous as in many parts of the Global South. Likewise, some of the elites found in Southern cities such as Mumbai and Sao Paulo would compare favourably with their Northern counterparts in London, New York and Tokyo regarding access to consumer goods and lifestyles.

However, these words of caution should not disguise the fact that profound economic and political divisions between North and South will persist well into this century, notwithstanding changes in particular countries and economies such as the East Asian tigers. For some sceptical commentators, the prospects for the Third World appear bleak because of four major factors: a reduction in aid and investment from the North to the South, a rise in racism and anti-immigration politics in the North, an increased tendency by powerful states to pressurize the South over debt rescheduling and trade access, and a reluctance on the part of the North to dismantle subsidy regimes which offer over $300 billion a year to Northern farmers alone. By way of contrast, the G8 offered only $8 billion in aid to Africa in 2001–2. For many commentators in the South, the current penchant for securing 'market access' to the world economy will ensure that Northern states continue to exploit the vulnerable and poorer zones. Although the rationale for the Cold War may have disappeared, the forces of economic globalization and supra-national capitalism will ensure that the power-geometries of North–South relations remain unequal and fractured.

Key questions

- What do geographical labels such as *global apartheid* suggest about the nature of globalization?
- Why have inequalities worsened between the North and the Global South?
- Why is capital supposed to flow freely and people not?
- What was the purpose of the Non-Aligned Movement? Does it still matter in a post-Cold War era?
- Why did the September 2003 WTO meeting end in apparently abject failure?

Further reading

For very good summaries of North–South relations and the Cold War see F. Halliday, *Cold War, Third World* (London, Verso, 1989), C. Thomas, *In Search of Security: The Third World in International Relations* (Brighton, Harvester, 1987). On non-alignment see P. Willetts, *The Non-Aligned Movement* (London, Pinter, 1978) and A. Singham and S. Hune, *Non-Alignment in an Age of Alignment* (London, Zed, 1986). On development see A. Escobar, *Encountering Development* (Princeton, Princeton University Press, 1995), G. Rist, *History of Development* (London, Zed, 1997) and D. Simon and K. Dodds (eds.), *Rethinking Geographies of Development*, special issue of *Third World Quarterly* 19 (4) 1998. On the condition of the former Soviet Union see M. Bradshaw and A. Stenning (eds.), *East Central Europe and the Former Soviet Union* (Harlow, Pearson Education, 2004).

Websites

Non-Aligned Movement www.nam.gov.org
Oxfam www.Oxfam.org.uk
UNDP www.undp.org

Chapter 4

POPULAR GEOPOLITICS

Key issues

- Why do the media matter in terms of shaping international politics?
- How is popular geopolitics linked to formal and practical geopolitics?
- How does media ownership influence the production and consumption of particular media?
- How do different types of media produce different types of popular geopolitics?

Images and other forms of representation of world politics are profoundly important in shaping patterns and responses to world political events. One of the defining images of the twenty-first century was created on 11 September 2001 when two planes flew into the World Trade Center in New York City (Fig. 4.1). The impact of the planes

Figure 4.1 The September 11th attacks on the World Trade Center in New York
Photo: PA Photos

and the subsequent diffusion of explosive flames throughout the buildings still haunt many who witnessed the event whether at Ground Zero or on television. This distressing series of images was endlessly repeated on virtually every television channel across the world. When some of the horrified onlookers were asked for their initial reactions to what they had just witnessed, a striking number cited films such as *The Towering Inferno* (1974) and even *The Siege* (1998), which addresses the question of 'Islamic terrorism' in New York. While CNN labelled the event part of 'America's New War', reality and fiction began to blur as witnesses tried to make sense of these momentous events.

The invocation of film in such a moment of shock and uncertainty should not necessarily surprise us. Most people in North America, Japan and Europe learn about foreign affairs through the media, whether through watching the television, listening to the radio or 'surfing' the Internet (see In focus 4.1). Indeed, three years on, the Google search engine reveals a staggering 8.9 million references to 'September 11th'. Developed by the US military in collaboration with US universities, the Internet has arguably revolutionized social life from the 1990s onwards. Over 500 million people are thought to routinely access the Internet for information and email. As a global network of information generation and exchange, however, it remains extremely unequal. The highest users of the Internet remain North Americans (35 per cent of global users), followed by Europeans (30 per cent). Less than 0.5 per cent of people in India have access to the Internet and the 'digital divide' is highly skewed in the Global South as urban dwellers dominate the usage figures. In Sub-Saharan Africa, radio (including the BBC World Service) remains the most important source of news information, as many people do not even have access to a telephone, let alone a computer.

It is probably not unreasonable to assume that in places such as North America and Europe television coverage played a key role in raising public awareness of recent humanitarian crises such as those in Rwanda (1994) and Kosovo (1999). Television and the Internet help re-frame places and locations (Fig. 4.2) by collecting, presenting and circulating information about the world. These technologies have enabled distant events and places to be transported via images and news stories into the homes of people living predominantly in the North. The transmission, circulation and reception of information and images is never a neutral process; places such as Bosnia and Iraq received considerably more television coverage in the 1990s than the civil wars and complex humanitarian emergencies in Angola, Kashmir and Chechnya. This has led many media observers to conclude that television coverage (and the large corporations which dominate global broadcasting) often unwittingly follows or helps to shape the foreign-policy agendas of powerful states such as the USA, France, Russia and the UK.

Not all televisual stories, newspaper features or films are simply propaganda. For our purposes, propaganda is defined as the deliberate construction and release of information which is designed to mislead or misrepresent particular situations and circumstances. We need to investigate how various sources construct particular interpretations of events, places or processes, such as the Cold War or the September 11th attacks, which in turn may influence specific courses of action. When witnesses to September 11th cited films as part of their explanation of such an event, it highlighted how popular culture and its conventions contribute to the context in which our ideas

Figure 4.2   CNN International pioneered the development of 24/7 television reporting
Photo: PA Photos

about people and places are framed and interpreted. This lies at the heart of what has been called **popular geopolitics**.

Popular geopolitics and the mass media

Before considering a range of popular geopolitical examples, it is worthwhile to explore how an interest in film or television might extend the remit of geopolitical research. Traditional geopolitics assumes that the geographical assumptions, designations and understandings of world politics are restricted either to the formal geopolitical models of well-known theorists such as Halford Mackinder or to the policy statements of national leaders and their political colleagues. The term 'popular geopolitics' is used to signify how political and media elites often attempt to represent the world and their position in consistent and regular ways (see Sharp 2000, McFarlane and Hay 2003). These representations may reinforce hegemonic ideologies such as transnational liberalism or, in the aftermath of September 11th, a strong association between terrorism and Islam. Thus, if one of the tasks of critical geopolitics is to challenge hegemonic representations of global politics then we need to be attentive to the interconnections with popular culture and the way in which newspapers and other media forms might either reinforce or contest geopolitical images and or representations.

Four working assumptions underwrite the emerging literature on popular geopolitics and associated media theory:

1. The media can play a key role in *agenda setting*. This means that media stories can help shape the ways in which particular events and or processes are represented and interpreted. They also, as part of that process, highlight some events/people/processes at the expense of others. During the 1990s, the humanitarian crisis in Bosnia was mainstream news while humanitarian emergencies in Sudan and Kashmir received less attention.
2. The media often *frame* events and processes and hence contribute to particular modes of interpretation or narrative structure. During the Yugoslav Crisis of the 1990s, for example, the British print media alternated between describing Bosnia as a 'holocaust' and as a 'quagmire'. The former helped to construct a clear sense of moral obligation to relieve human suffering (given the associations with the Nazi genocide of Jewish communities) while the latter implied that intervention might be hopeless and costly (with allusions to the First World War and or the American involvement in Vietnam).
3. Film and other media forms can be used to explore how commonsense stereotyping of Others (inside and outside a particular state) contributes to the articulation and reproduction of *national identities*. Cold War cinema provides a particularly rich example, as American film companies produced films that reinforced an image of the Soviet Union as the Evil Other. However, film companies produced films which subverted such an assumption about the geographical location of threat.
4. The consumption of films and other forms of media is not always obvious or in a manner intended by the producers and or sponsors. There is scope for a *multiplicity of interpretations* of media material. While films such as *Top Gun* (1986) and *Iron Eagle* (1985) were, in my view, deeply sympathetic to the Cold War agendas of the Reagan administration, many viewers may simply have been oblivious to such geopolitical connections. Just like the media generally, audiences (around the world and not just in Britain or America) can also subvert as well as reinforce and support particular geopolitical visions.

In contrast to many mainstream realist and liberal approaches to world politics (**formal geopolitics**), critical geopolitical authors have argued that ideas and representations about the political world are expressed and reproduced outside the narrow confines of the diplomatic circuit, foreign-policy decision-making and intergovernmental conferences (**practical geopolitics**). In other words, the diplomatic conference room and the battlefield are not considered disconnected and/or divorced from public culture (see In focus 4.1 and Fig. 4.3).

As established in earlier chapters, geopolitics is considered to be a series of problematics concerning power, knowledge, space and identity. We have already explored, for example, how geopolitics can be considered as a particular discourse on statecraft and state power and how and with what consequences particular networks of power-knowledge construct hegemonic geographical and political identities. An examination

In focus 4.1: Territory and public culture: the case of Argentina and the Islas Malvinas

In Argentina, stories and representations of the Islas Malvinas (Falkland Islands) are to be found in murals, postage stamps, atlases, monuments, popular songs, countless newspaper articles and television shows. Underlying these representations remains a widely held belief that Britain illegally colonized these Argentine islands in the South West Atlantic. Since the 1830s, successive Argentine governments have not only protested against this occupation but have also sought to remind their citizens that Argentina remains a geographically incomplete nation until the reclamation of the Islas Malvinas. President Juan Perón ordered the Military Institute of Geography to impose a new law demanding that all maps of Argentina represented the Malvinas as an Argentine rather than British de facto territory. By raising public awareness, it becomes perhaps more understandable how and why an unpopular Argentine military regime could gather substantial popular support for the invasion of the Falklands in April 1982. While the official visit of President Menem of Argentina to the UK in October 1998 (the first by an Argentine president since 1961) helped to foster more cordial relations, successive Argentine presidents have committed themselves to pursuing this particular territorial claim.

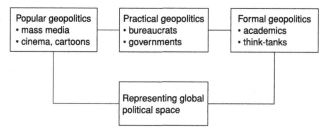

Figure 4.3   Linking formal, popular and practical geopolitics
Source: Adapted from O Tuathail and Dalby 1998

of popular sources such as television and magazines offers critical geopolitics ample possibilities for discerning how other places and peoples are represented within a variety of national and cultural contexts. Five such 'popular' sources will be considered here: films, television, the popular magazine *Reader's Digest*, cartoons and music. Another important source, the Internet, is considered in later chapters examining the anti-globalization movement and humanitarianism (Chapters 7 and 8). All contribute in an interrelated manner to the construction of geopolitical life-worlds of citizens of states and to the wider global polity.

Films and geopolitical visions

The film industry played a very important role in the twentieth century. It contributed greatly to debates over historical accuracy, political agendas and dominant social tendencies (Davies and Wells 2002: 5). While the cinema provides mass entertainment and story telling, the effects of film as a medium need to be recognized. Governments and their militaries have collaborated closely with film companies, presumably because they believed that film could shape messages and meanings of events. Even when film makers were not actively collaborating with political leaderships, film genres such as the Western (e.g. *The Good, the Bad and the Ugly*, 1965) can be interpreted as highly political in the manner by which they seek to represent 'frontier America', individual heroism and the struggle between good and evil (Short 1991: 178–96). As Davies and Wells conclude, 'The fact remains, though, that the politics of the most powerful nation in the world [the USA] cannot be divorced from the most far-reaching entertainment medium in the world' (2002: 5).

As a technology, film is ideally suited for projecting political, social, moral and cultural views to audiences around the world. American and European cinema has had a long and complex relationship with political institutions, public opinion and national identity formation. D. W. Griffith's film *The Birth of a Nation* (1915) was highly significant in constructing a particular narrative about American national identity. As O Tuathail noted, '*The Birth of a Nation* portrays the Ku Klux Klan as the saviours of the white race, as defenders of the virtue of white women, and as representatives of the Christian civilization, a civilization under threat from the innate primitism [sic] and uncontrollable sexual appetite of emancipated African-Americans' (O Tuathail 1994: 540).

Early silent films were very popular with American and other European audiences. In the enclosed atmosphere of the film house, cinema had a tremendous potential to hold the attention of the audience. Novel forms of visual presentation were quickly adopted by political and media elites alike, as a powerful propaganda tool which could be used to screen epic tales of nation formation and identity politics. Under Joseph Stalin, for example, Soviet cinema flourished in the 1920s and 1930s. The Soviet Communist Party funded cinematic projects such as *October* (1927) followed by a stream of films in the 1930s such as *Two Captains* and documentaries depicting the Arctic exploits of Soviet pilots (McCannon 1998). The 'myth of the Arctic' became a central feature of Stalinist popular culture as Soviet citizens were cajoled into taking an interest in the exploration and 'conquest' of the North Pole. The purpose of such financial and cultural invest-ment was to demonstrate that the communist state could conquer any obstacle placed in front of it, natural or unnatural. These films helped to construct particular polit-ical and cultural identities linked to national prestige and socio-economic development. In the aftermath of the Second World War, political and cultural repression were the defining norms and only films that focused on approved subjects such as the 'heroic' role of the Soviet Union during the war were produced. Other productions approved by the Stalinist Soviet Union included historical epics such as the two-part *Ivan the Terrible* (1942 and 1946) and/or anti-American films such as *Court of Honour* (1949), which warned citizens about the seductive dangers posed by US society and its conspicuous consumption (Dodds 2003a).

In the United States, it was often noted that wars tended to be fought twice: first on the battlefield and then on celluloid. The relationship between governments and film production companies was often intimate given the enormous potential for influencing public opinion. In the first half of the twentieth century, it was common for governments including most famously that of Hitler's Germany to co-operate very closely with film makers on specific projects. Over the last eighty years, there have been countless examples of the American government co-operating closely with Hollywood. The former president Theodore Roosevelt asked the Wilson administration to approve the release of 2,500 marines for the making of the picture *The Battle Cry of Peace* (1915), which concentrated on exposing the fragile nature of America's armed forces at the beginning of the First World War. It was mooted that the popularity of this film helped to persuade the American president, Woodrow Wilson, not only to build up America's fighting capacity but also to enter the First World War in 1917.

After the military disaster at Pearl Harbor in 1941, Franklin Roosevelt's administration approved the lease of numerous planes and ships in order to produce the movie *Air Force*, which was released for general viewing in 1943. The film was intended to reassure Americans that the country would be able to combat Japanese attacks on domestic territory. At the end of the film, the American air force was shown (two years prematurely) to have triumphed over the Japanese war machine. In a similar fashion, Walt Disney commissioned the production of *Victory Through Air Power* (1942) in order to demonstrate to American citizens the strategic significance of aircraft and their role in the American war effort against Japan and Germany. The American director Frank Capra was also instrumental in this role as executive producer in 1942 and 1943 of the *Why We Fight* films, which were designed to strengthen American troop morale and to explain the reasons for American involvement in the conflict with Japan. During the Second World War, Capra commanded the 834th Signal Division of the US Army; he later used Nazi propaganda films to construct anti-fascist narratives and films such as *Prelude to War*. Unsurprisingly, the release of these films coincided with a rapid increase in the internment of Japanese-American citizens in the USA (see O Tuathail 1994: 541).

In the post-1945 period, Hollywood continued the comfortable relationship with the Pentagon. During this period, the US military established a public-relations office in Beverley Hills, Los Angeles in order to consider producers' requests for equipment and special assistance. Two types of film projects which appeared to engender considerable support from successive American administrations concerned the Second World War and the post-1945 threat posed by the Soviet Union. The parallels with post-1945 development of Soviet cinema are thus considerable as American film producers also experienced cultural suffocation. In 1947 the US House of Representatives Un-American Activities Committee investigated Hollywood and 'black-listed' writers, producers and actors for their alleged connections to communism and un-American activities. Former B-movie actor and later president of the United States Ronald Reagan played his part in helping to identify 'subversive' elements in Hollywood.

In the approved war film, *The Longest Day* (1961), the Defense Department gave permission for around 700 troops to be used as film extras during Darryl Zanuck's reconstruction of the Normandy beach landings in 1944. However, due to the developing Berlin crisis in 1961, Secretary of State Robert MacNamara ordered that this film contingent be cut to 250 personnel in the light of fears that the Soviet Union was about to invade the Western sectors of the city.

In the ideological struggle against the Soviet Union, the Eisenhower government (1952–60) gave permission for the CIA to covertly support the production process of the first feature-length British animation film based on George Orwell's *Animal Farm* (see below under 'Cartoons' and Fig. 4.4). Under the direction of John Halas and Joy Batchelor, filming began in 1951 and was completed by 1954. The original idea for the film came from the American film producer Louis de Rochemont, who was linked to the CIA-funded Congress for Cultural Freedom, which was designed to produce and circulate material sympathetic to the American 'way of life' based on democracy, market economics and the freedom of choice (see Whitfield 1991). The involvement of the CIA in the actual production process was mainly concerned with the ending of the film. In the original story by George Orwell, the triumphant animal revolutionaries (in the main the pigs) become the new elite and behave in a manner reminiscent of their human masters. The book ends on a pessimistic note with the other animals (such as horses, chickens and cows) meekly accepting their new conditions without active resistance. In the film version, the animal leaders (Napoleon and his dogs) are overthrown by the other animals, who are disaffected with their corrupt ways of governance. For the Psychological Strategy Board of the CIA and the US National Security Council, this ending of the film was crucial because it demonstrated that new forms of oppression (read the post-war Soviet Union) could be overthrown if the oppressed (read Eastern European states such as Czechoslovakia, Poland and Hungary) were organized and prepared to revolt.

The relationship between Hollywood and the US military became more problematic during the 1960s, a decade dominated by Vietnam and the US civil rights movement. The only significant film of the conflict made by US producers during the 1960s was *The Green Berets* (1968), starring and directed by John Wayne. However, even his presence as lead actor failed to allay concern among some senior US military officers about the film's portrayal of the violent struggle in the jungles of South East Asia. The timing of the film was controversial too, because President Johnson had already admitted that the war was a 'bitch' that had caused considerable damage to the prestige and reputation of the US around the world. In the film a Green Beret soldier lectures a journalist about the need to resist 'the intentional murder and torture of innocent women and children by the communists . . . I tell you these people need us, they want us' (cited in Pilger 1998: 561). The film grossed $8 million in 1968–9 and was considered to be a morale-boosting movie in the wake of the 1968 My Lai massacre which entailed the killing of hundreds of unarmed Vietnamese men, women and children by American troops led by Lieutenant William Calley.

A decade later, film makers like Francis Coppola were denied co-operation by the American Defense Department, and the film *Apocalypse Now* (1979) was produced in

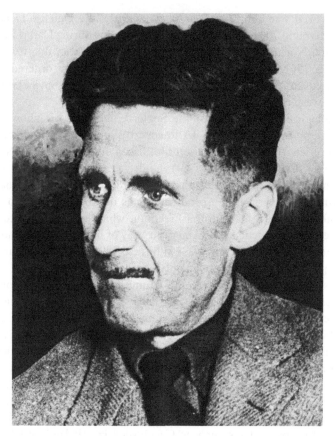

Figure 4.4   George Orwell
Photo: © Bettmann/CORBIS

the Philippines with the assistance of the Philippine armed forces. Ironically, the infamous helicopter attack scenes (accompanied by the music of Richard Wagner) were nearly cancelled, as President Ferdinand Marcos demanded the return of the helicopters because they were urgently needed for some 'real' military action in another part of the islands. During the filming process, the main actors and the director were profoundly affected by the experience of recreating the combat scenes and the search for the rogue Colonel Kurtz.

The phrase 'Vietnam Syndrome' was coined in the 1970s in response to fears among American political elites that the humiliation in South East Asia had caused widespread feelings of depression, guilt and a loss of moral purpose. Over 58,000 American servicemen had lost their lives in a decade of fighting and some are still considered Missing in Action (MIAs). This led to a reluctance for further involvement in the affairs of poorly understood and distant places. After the Vietnam conflict, Hollywood produced numerous films, such as *Hamburger Hill* (1985), *Platoon* (1986) and *Born on the 4th of July* (1995), which sought to sympathetically portray the harrowing experiences of American soldiers in Vietnam and on their return to the USA. The popularity of these films contrasts with the reaction of mainstream American society, which shunned returning Vietnam veterans in the 1970s. This ambivalence was evident in the 1982 film *First Blood* starring Sylvester Stallone as a psychologically disturbed Vietnam veteran persecuted by a malicious police chief in the North West of the United States. Forced to rely on his Special Forces training, John Rambo successfully outwits the local police and state guard until his eventual surrender to his former Vietnam military commander. The final stages of the film offer no clues as to the fate awaiting a former serviceman clearly in need of medical support.

During the final phase of the Cold War, the movie *Top Gun* (1986) featuring a successful US navy pilot (code name Maverick) was actively supported by the American armed forces as it was seen to present a positive image of the navy and its aviators. The American navy supplied a number of F-15 aeroplanes and the aircraft carrier *Enterprise* for the duration of filming, and by the end of 1986 the film had earned $130 million, eventually grossing $350 million in worldwide cinema sales (Kellner 1995: 80). The media critic Douglas Kellner has argued that *Top Gun* was indicative of a particular period of American foreign policy characterised by

> aggressive military intervention in the Third World, with an invasion of Grenada, the US-directed and financed Contra war against Nicaragua, the bombing of Libya, and many other secret wars and covert operations around the globe. Hollywood films nurtured this militarist mindset and thus provided cultural representations that mobilised support for such aggressive policy . . . the 'enemy' [in *Top Gun*] flies MIGs, a Soviet plane, but is not identified as Russian [sic], though the MIG fighter pilots have red stars on their helmets (Kellner 1995: 75).

This interpretation of *Top Gun* is widely shared by media scholars, amongst them Shohat and Stam, who have argued that this film along with others such as *Rambo* epitomized the Reagan administration (1981–9) ethos of militarism, hostility to the Soviet Union, anti-intellectualism and social conservatism (see Shohat and Stam 1994). Moreover, Susan Jeffords has contended that Hollywood's fascination with the so-called 'hard bodies' of figures such as Rambo not only epitomized the character of 1980s America but also served to highlight a difference between the Reagan and the former Carter administration which had been unable (in a humiliating way) to rescue the American hostages held at the American Embassy in Tehran (Jeffords 1994).

In contrast, *Courage Under Fire* (1996) did not enjoy the military assistance given to the producers of *Top Gun*. The request for the lease of M1 Abrahams tanks, Bradley fighting vehicles and Blackhawk helicopters was not met by the US armed forces. Media analysts considered this film noteworthy because it was the first American production to deal with the country's participation in Operation Desert Storm during 1990 and 1991, a UN-sponsored campaign which witnessed the triumph of the United States and its allies over the Iraqi forces of Saddam Hussein. The American military authorities responsible for co-operation with Hollywood productions rejected the call for assistance because the film dealt with the controversial issue of 'friendly fire' and attempts by senior officers to cover this up during the campaign. There were also demands that the main character (the black actor Denzil Washington) should not appear drunk whilst wearing a military uniform. When the producers refused to alter their script to the extent demanded by the US military, they were forced to import 12 British tanks and two Cobra helicopters in order to reconstruct the battle scenes.

Films can be a very rich and varied source for political geographers as well as film critics due to the widespread distribution of particular productions and the massive audience potential. Debates about the influence and the connections between image and real-life international political behaviour continue, often focusing on the manner in which films depict certain individuals or groups (e.g. Muslims) as threats to or enemies of the United States. In terms of international affairs, there have been many examples of governments and leaders using the cinema to manipulate public opinion, often in times of crisis or war. The interactions between governments and film-producing centres have been substantial and at times prone to subtle interference in the final production, as the examples above have illustrated. These connections are meaningful because the US military will not lend equipment such as planes and tanks to film projects which it considers unsympathetic to the armed forces. Sometimes particular representations of war and 'threats' coincide with real-life events, as in the case of the film *Black Rain* (starring Michael Douglas, Andy Garcia and Ken Takakura), which was released in 1990 at a time when America was locked into a bitter trade war with Japan. Filmed in Japan, the screenplay concentrates on the struggles of two US detectives attempting to arrest a Japanese mafia figure amongst the violence and strangeness of a gangland-riddled Japanese society. The uneasy relationship between America and Japan is a central theme in the film; the 'black rain' of the title refers to the fallout from the nuclear bombs which fell on Hiroshima and Nagasaki in 1945 (Morley and Robins 1995: 161).

Released in November 1998, *The Siege* opened to considerable controversy in the United States because it featured 'Islamic terrorists' operating in Brooklyn, New York, subsequently pursued by a powerful agent (played by Bruce Willis) who was appointed by the US government to crush the bombing operations. The Council of American-Islamic Relations (CAIR) and American-Arab groups complained that the images of terror perpetuated ethnic and place-based stereotypes about Islam and the Middle East. In the light of the ongoing negotiations between Arabs and Jews over the future of Israel and the West Bank, Palestinian commentators such as the late Edward Said (Professor of Comparative Literature at Columbia University in New York) have warned that these representations of Muslims and Arabs influence public attitudes towards real-life political negotiations. This concern was reiterated following the release of *Rules of Engagement* (2000), which depicted US Marines killing civilians protesting in a square below the American Embassy in the Yemen. Supported by the Pentagon, the film shows the accused Marines being exonerated when it is demonstrated that many in the crowd, including children, fired on them: the film concludes that the Marines were responding to an unprovoked attack rather than slaughtering unarmed civilians. The American-Arab Anti-Discrimination Committee, whose members reported that many white American cinema viewers cheered when the Marines killed members of the crowd, condemned the film (see In focus 4.2).

In focus 4.2: Hollywood and the post-Cold War era

When the Cold War ended in the late 1980s, Hollywood (alongside the US government) faced a dilemma. What new dangers confronted the US given the collapse of the Soviet Union? In order to engage audiences with a storyline, many action/thriller films base the main narratives (however loosely) on contemporary geopolitical events. With the end of the Cold War, Hollywood had an opportunity to develop and/or reinforce new plots.

These new story frames have included nostalgic representations of the Second World War (*Pearl Harbour*, 2001) and the Cold War including the US-Soviet space race (*Apollo 13*, 1995), ex-Soviet terrorists (*Air Force One*, 1997), alien invasion (*Independence Day*, 1996), disease (*Outbreak*, 1995) and weapons of mass destruction (*The Rock*, 1996). Strikingly, Hollywood has also continued to produce films which depict Muslims and Arabs as a threat to the United States, a trend set in the 1980s by movies such as *Iron Eagle* (1985) and *Navy SEALS* (1986). In most cases, Muslim combatants are shown to be shadowy and yet predictable because of their predilection for violence and hostility towards the West. Such depictions arguably avoid harder ethical reflection concerning the geopolitical role of the United States and its allies in the Middle East and Islamic world (see Chapter 9).

A word of caution is due when considering the interpretation of films and their possible cultural and political influence (even though analysis of audience reaction may well be possible through audience surveys and film media critiques): there is no guarantee that the viewing public will adopt the meanings the directors and politicians have anticipated. For example, in *Top Gun* it is possible to perceive a range of scenarios. The aircraft used in the attack scenes appear to be MIG jets but were actually American Northrop Freedom Fighters, and the enemy's red stars need not necessarily be Soviet but could refer to to various Arab and or Chinese armed forces. The red star is highly symbolic because it is identified as referring to communist countries and socialist ideologies. Yet *Top Gun* could be interpreted as a film supportive of the aggressive military action of the United States in the Middle East during the 1980s culminating in the deployment of troops to the Lebanon. Alternatively, many viewers of *Top Gun* may not have made those connections at all and simply thought it was an action-packed film about planes, pilots and the personal relationship between 'Maverick' and his flight instructor, Charlie (played by Kelly McGillis).

Film and cinema therefore offer exciting possibilities for considering the prevalent geopolitical representations of world politics and places (see In focus 4.3 and Fig. 4.5). The popular movie cultures generated by Hollywood clearly have a tremendous impact in terms of audience figures and revenue generation. However, the interpretation of film (and, for example, the analysis of 'threat' construction) needs to be approached with caution, as there is no automatic or causal link between the film and audience reaction.

In focus 4.3: James Bond: British icon? Western icon? Global icon?

Created by the former journalist and wartime spy Ian Fleming, the superspy James Bond/007 has arguably emerged as Britain's greatest contribution to modern genre cinema. With the backing of United Artists, Bond's first outing, *Dr No* (1962), saw him confront the evil genius Dr No and his plans to disrupt US space operations in the Caribbean. Released in the midst of the Cuban Missile Crisis, it had a geopolitical relevance that audiences around the world were able to recognize. Despite operating in the US's geopolitical 'backyard', it was a British rather than an American agent who was able to save the world from a potential crisis. For British audiences, Bond (initially played by the Scottish actor Sean Connery) represented a rather appealing post-imperial hero. In a decade of continued colonial losses, and political subservience to the superpowers, Bond was the senior partner in the Anglo-American relationship and travelled around the world foiling plots by the Soviets and the non-aligned criminal organization, SPECTRE.

When President John Kennedy admitted that Ian Fleming was one of his favourite authors, Bond's popularity as a Western icon was confirmed. Fleming was even invited by Kennedy and his brother Robert to advise them on how 'James Bond' might topple the communist Cuban leader Fidel Castro. Fleming suggested a variety of ideas, including poisoned cigars, exploding seashells and even dropping leaflets over Cuba claiming that Castro was impotent. Bond, like the Kennedy brothers, became not only an icon of style but also representative of a hyper-masculinity in an era when feminism challenged Western patriarchy.

By the late 1960s the Bond films were established as a global cinematic phenomenon. The Bond formula of an attractive lead actor, fast action, exotic locations, glamorous women and gadgets was said to be responsible for the extraordinary commercial success. In order to keep the formula fresh, the films have also deliberately moved with the geopolitical times. In the midst of the Cold War, films such as *You Only Live Twice* (1967) and *For Your Eyes Only* (1981) developed stories involving an East–West confrontation. In a period of relative détente Bond is seen working with his Soviet counterpart (*The Spy Who Loved Me*, 1977) and post-Cold War films such as *Goldeneye* (1995), *The World Is Not Enough* (1999) and *Die Another Day* (2002) have deliberately been located in the fragmenting Soviet Union and the 'axis of evil' state of North Korea. *Tomorrow Never Dies* (1997) considered the role media moguls could play in engineering a major confrontation between Britain and China for the purpose of securing exclusive media rights in the Chinese market.

After 20 films, there is no evidence to suggest that Bond's popularity has diminished and plans are afoot to shoot a new Bond film in 2005. It is estimated that at least 25 per cent of the world's population has seen one Bond film or more.

Further reading: *see* Black 2001.

Figure 4.5   James Bond in the new millennium, played here by the actor Pierce Brosnan
Photo: PA Photos

News media and the 'CNN factor'

The last two decades have witnessed the growth of globally orientated events and the increasingly high-level coverage of war and humanitarian disasters such as the 1999 NATO bombing of Kosovo and the US coalition invasions of Afghanistan (2001) and Iraq (2003). Television provides a potentially revolutionary medium for information exchange because of its ability to transcend spatial and social boundaries around the world. As such it helps to *agenda set* and *frame* events for elite and popular audiences. This matters because the way in which stories are constructed and then presented may, in the words of Edward Herman and Noam Chomsky, help in the process of 'manufacturing consent' for particular foreign policy agendas of national governments (Herman and Chomsky 1988).

The relationship between television coverage and international politics has recently attracted much critical attention in media studies, international politics and geopolitics (Gowing 1994, Robinson 1999, 2002). Major media networks such as Cable News Network (CNN), AOL and Time-Warner, News Corporation and Sony have demonstrated a remarkable capacity for informing or even shaping international agendas, leading some analysts to talk of a 'CNN factor' (see Robinson 2002 and In focus 4.4).

The advent of portable video cameras, satellite dishes and freelance journalism has further contributed to this phenomenal growth and it is estimated that around 200 television channels could be licensed following further attempts to loosen media ownership rules (Morley and Robins 1995: 13). In conjunction with other media empires such as Time-Warner and News International, CNN occupies a powerful position in terms of audience figures and news production within the global broadcasting world.

Globalization theorists often refer to the rise of real-time television networks as further evidence of time–space compression as the experiences of distant places are brought into the living rooms of (in the main) Western citizens. The BBC's coverage

---

In focus 4.4: The CNN factor

The CNN factor refers to the 24-hour Cable News Network channel based in Atlanta, Georgia (USA), which has provided worldwide English-language news since 1980. It rose to global prominence with live coverage of the 1991 Persian Gulf War. Few will forget the reporting of Peter Arnett in the midst of an American attack on Iraqi positions just hours after the deadline for Iraq to withdraw from Kuwait. It was later claimed that even the then US president George Bush Senior was avidly watching the CNN coverage in order to keep abreast of a rapidly changing situation. His opponent, Saddam Hussein, invited CNN to conduct interviews with himself and even some British hostages held in Baghdad.

More generally, in the aftermath of the 1991 Persian Gulf War it has been alleged that television coverage of humanitarian disasters can cajole Western democracies into taking political action because of fears of adverse reactions from voters. The evidence for this causal link is extremely patchy. Despite countless reports of genocidal violence in Rwanda in April 1994, no Western government sought to intervene to prevent further loss of life. Even the much-cited television coverage of Somalia in 1992 occurred only after a political decision had been taken to send US troops on a UN-sponsored mission to improve humanitarian aid to that war-torn country. Bosnians suffered three years of intense humanitarian suffering before the US-led NATO force decisively intervened in 1995 to prevent further loss of life. Images of genocidal violence, mass rape and destruction had been shown on Western television continuously since 1992. Generally, it appears that political will is far more important than television coverage.

of starving children during the 1984 Ethiopian famine was believed to have been the impetus for television viewers to contribute to the relief appeals. The song 'Do they know it's Christmas', released by a group of artists under the label of Band Aid, became one of the biggest-selling singles in 1984 following a pledge that all proceeds from its sale would be donated to famine relief in Ethiopia. Television coverage of tragic events such as famine and war can obscure geopolitical issues as events are transformed and re-presented. In the case of Ethiopia, the news of a massive famine coincided with the intensification of the Cold War between the United States and the Soviet Union in various areas of the world including Sub-Saharan Africa. Viewers were presented with ever more distressing pictures of starving children yet relatively little coverage was devoted to wider issues concerning the origins of the famine such as the political and economic context of the country. This humanitarian disaster was not entirely due to the failure of the rains but was undoubtedly exacerbated by the ongoing civil war between rival factions (supported by either the USSR or the USA) and the fact that the fertile lands of Ethiopia were being used to grow cotton for export, revenue from which was used by the ruling government to purchase further weapons (see O'Loughlin 1989). Apparently after the American NBC channel covered the famine story for American viewers, the aid given to Ethiopia by the Reagan government increased from $20 million to $100 million in the course of 1984 (Harrison and Palmer 1986).

The experience of the 1984 Ethiopian famine crisis and other notable events raised a number of controversial issues with regard to the role of television and its influence on political decision-making and public opinion. The first problematic was advanced by the French philosopher Jean Baudrillard, who published a series of provocative articles in the French newspaper *Libération* from January 1991 onwards. In the case of one particularly famous article entitled 'The Gulf War did not happen', Baudrillard argued that the massive increase in television images over the last 30 years has changed our relationship to the wider world. More specifically, he contended that television coverage of the 1991 Persian Gulf War was increasingly comparable to a movie or a series of images rather than footage of real-life events such as the bombing of Iraqi troop positions, power stations in Baghdad and other civilian and military installations in Iraq. In an often misunderstood argument, Baudrillard is not suggesting that Operation Desert Storm (i.e. the Allied assault on Iraq) did not happen but rather that the video images of the 1991 Persian Gulf campaign became more significant than the real-life events. Viewers were increasingly encouraged to watch images of the bombardment and compare video stills rather than to actively contemplate either the horrific consequences for human life of so-called smart bombs and cruise missiles or why Iraq was being invaded in the first place.

Television coverage of international politics, although capable of exposing injustice and mobilizing public opposition to brutal state violence, does not always provoke action that many viewers may have anticipated. Television pictures of the Chinese student protests in Tiananmen Square and the resultant massacre by the Chinese armed forces in June 1989 did not provoke other governments to take preventive or retaliatory action. In that sense Bernard Cohen's comments about the value of media reporting in Somalia during the 1992 UN operation overstates the case:

Television has demonstrated its power to move governments. By focusing daily on the starving children in Somalia, a pictoral story tailor-made for television, TV mobilised the conscience of the nation's public institutions, compelling the government [i.e. the US administration] into a policy of intervention for humanitarian reasons (cited in Mermin 1997: 385).

More recent research on the role television coverage plays in shaping American foreign policy has concluded that 'Somalia' (the scene of an UN operation in 1992) had already become a humanitarian issue within the foreign policy community in Washington before the camera crews arrived in Africa.

The 2003 US invasion of Iraq provides a more sinister example of how television coverage can be deeply constrained by commercial and geopolitical pressures in the aftermath of September 11th. The political economy surrounding media ownership, especially in the United States, must also be recognized. In 2003, nine corporations dominated US and global media; these included AOL Time Warner, Seagram, Sony, Disney and News Corporation. They collectively represent a massive and well-funded political lobby, which campaigns hard for greater commercial freedom. In the United States following September 11th and the introduction of new legislation such as the

Figure 4.6 Osama bin Laden used the Arab-language news channel Al-Jazeera to broadcast to the world
Photo: PA Photos

2001 Patriot Act (which allows the federal government to enhance surveillance of its citizens if national security is judged to be imperilled), there is mounting concern that the media are becoming increasingly politically conservative and simply not reporting events or individuals who are considered critics of the 'war on terror', for example. Many Americans (including high-profile Hollywood film stars such as Sean Penn, Susan Sarandon and Tim Robbins) have complained that they have been forced to access Internet-based versions of other English-language newspapers and or satellite television channels such as BBC World and/or Al-Jazeera to discover alternative viewpoints surrounding the US-led 2003 invasion of Iraq (Fig. 4.6).

The global reach of US media corporations often remains disturbingly parochial and politically conservative (see In focus 4.5).

There is ample evidence that the mainstream media in the US choose simply not to report anti-war protestors and that peace groups were refused the opportunity to purchase commercial airtime to contest the Bush pro-war strategy in 2002 and 2003. High-profile actors who opposed the war on Iraq such as Martin Sheen, Susan Sarandon and Sean Penn were accused of being un-American, and television companies reputedly warned them and other actors not to voice their opposition to the 2003 Iraq campaign because of fears that commercial sponsors would not wish to be

---

In focus 4.5: Scooping the US media? the rise of the Al-Jazeera satellite channel

Based in the Qatari capital Doha, the Al-Jazeera satellite channel (created in 1996) became globally prominent following the September 11th attacks on the United States. It was the first Arab news channel to broadcast videotapes featuring the speeches of the perpetrator of the attacks, Osama Bin Laden and the Al-Qaeda network. Despite US protests, the news channel continued to broadcast extracts from Al-Qaeda, often infuriating American officials who hoped that Osama Bin Laden had been killed following the invasion and bombing of Afghanistan in November 2001. The US military later 'accidentally' bombed the Al-Jazeera office in Kabul in Afghanistan.

Unusually for the Middle East, Al-Jazeera is not subject to government censorship and thus has earned a reputation throughout the region and the Arab Diaspora for its controversial style of news reporting. Talk-show programmes such as *The Opposite Direction* routinely debate controversial issues in Islam and the wider Islamic world, such as gender equality. Unlike Western broadcasters, Al-Jazeera also broadcast live television pictures of American and Iraqi casualties during the 2003 US-led invasion of Iraq. As a consequence viewers were asked to confront directly the very real human cost of warfare rather than avoid such shocking images in the name of protecting the feelings of relatives of the dead or national morale more generally.

Notwithstanding criticisms of Al-Jazeera's political agenda (it is not very critical of its government host, Qatar), it is the only Arab-language television network that unquestionably rivals the coverage provided by British, American and other Western media corporations, especially in the Middle East and Islamic world (see El-Nawawy and Iskandar 2002).

associated with 'anti–US' activism. In the midst of this controversy, the MSNBC cable news channel dropped liberal broadcaster Phil Donahue, citing declining viewing figures. Some contend that he was dropped because he was perceived to be too sympathetic to the anti-war protestors. While American media corporations and their news channels were avoiding the anti-war critics, lavish coverage was devoted to President Bush's '*Top Gun*' moment as he landed on board the USS *Abraham Lincoln* to congratulate the returning US navy in May 2003. It has been estimated that $1 million was spent on securing the right kind of television coverage.

The *Reader's Digest* and the Cold War

Popular magazines, like film and other forms of print media such as newspapers, encourage the reader to interpret and identify with issues ranging from consumer products to international political stories. While individual readers may well read those stories in differing manners, it is often the case that a particularly dominant interpretation or reading will emerge as hegemonic depending on the subject matter. During the Cold War, for example, American magazines often employed a mixture of tropes, frames and agendas based on individualism, consumer choice, morality and manifest destiny in order to promote geopolitical visions of 'America', often at the expense of other states and ideologies such as the Soviet Union and communism. They also used (and continue to use) an amalgam of pictures, maps, human-interest stories and background information to provide overviews of contemporary events and figures.

Popular magazines and journals such as the *Reader's Digest*, *Life* and the *National Geographic* have been long-standing features of Euro-American public life in the twentieth century. This assessment is based on a variety of criteria ranging from circulation figures to the political and cultural significance of the contributing authors. The *Reader's Digest*, for instance, enjoys a circulation of 16 million in the United States and is widely translated from English into a variety of other European and non-European languages (see Fig. 4.7).

The founder of the magazine, DeWitt Wallace, had a particular geopolitical vision, which was deeply sceptical of the Soviet Union, trade unions, totalitarian governance and communist politics. The Italian version of the *Reader's Digest* was launched in 1948, when it was feared that the Italian Communist Party would seize power following democratic elections (Sharp 1996: 567). A remarkable feature of the magazine was the ideological transformation of the Soviet Union from wartime ally to post-war adversary.

The British geographer Joanne Sharp has conclusively demonstrated how the *Reader's Digest* helped to shape American views of the Cold War through its selection and editorial presentation of articles and stories (Sharp 1993, 1996, 2000). From the 1920s onwards, the unsettling and potentially threatening character of the Soviet Union generated a steady stream of articles on the totalitarian nature of Soviet political life. Sharp argues that the US version of the magazine participated in maintaining a particularly dominant representation of the Soviet Union throughout the Cold War

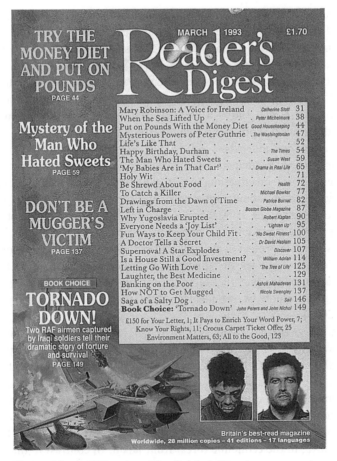

Figure 4.7   A 1993 cover of the *Reader's Digest*
Source: Reprinted by permission of *Reader's Digest*

period and actively encouraged the reader to identify with American interests at the expense of the 'Soviets' or 'Reds', who were portrayed as a threat to the 'American way of life'. Other geographers have also noted these kinds of geopolitical strategies:

> Now readers were bombarded with accounts of the total difference of the opposing character of the Soviet Union to the 'actually existing' United States, not simply differences in aspiration or historical experience. Such representations had powerful effects in mobilising public opinion into the Cold War consensus that progressively engulfed American politics from 1947 until the Vietnam War of the late 1960s (Agnew 1998: 110).

On the domestic front, the *Reader's Digest* portrayed a particularly narrow vision of American political identity based on the belief that the American economy and society must be supported *inter alia* by mass consumption and production, an enlarged military presence in the wider world and complete hostility to any form of left-wing or labour militancy. As the American political scene of the 1950s and 1960s demonstrated, free and competitive democratic politics was not possible in a climate of profound paranoia towards socialism and communism. As O Tuathail and Agnew concluded, 'The simple story of a great struggle between a democratic "West" against a formidable and expansionist "East" became the most dominant and durable geopolitical script of the [Cold War] period' (1992: 190) – a narrative that was reproduced in countless forms of media and cultural expressions ranging from films and television broadcasts to popular magazines and journals.

The *Reader's Digest* participated in the construction and legitimation of a particular Cold War geopolitical imagination, which witnessed the division of the world into competing blocs. Stories about communism and the 'Soviet way of life' were counter-posed with representations of 'America' which tended to emphasize not only national territory but also a transcendental morality according to which the United States had a manifest duty to protect democracy and capitalism for the wider world. In a trend that echoed earlier episodes of American exceptionalism, 'America' became transformed into a symbolic space ('a beacon on the hill'), which had a moral responsibility to protect and promote its own ideological and material vision for the post-war world (see Agnew 2003).

Ironically, the *Reader's Digest* was extremely sceptical of Mikhail Gorbachev's reformist government in the Soviet Union during the mid-1980s (Sharp 1996). In the period after the Cold War, the magazine continued to warn its readers that the Soviet Union might not be capable of change in terms of totalitarianism and the apparent rejection of market economies. However, it also raised the possibility that America was now threatened with other dangers ranging from domestic terrorism (such as the 1995 Oklahoma bombing by far-right fanatics), to Islamic fundamentalism in places such as Iran and Libya, to the economic threat from Japan. In contrast to the Cold War, the geographies of danger appeared varied and complex: the reader of *Reader's Digest* in the 1990s was warned that threats lay within and outside the boundaries of the United States (Sharp 2000). Following the September 11th attacks on the United States, this sense of domestic insecurity has been heightened.

Cartoons and the anti-geopolitical eye

Political cartoons have a long and varied political and artistic history ranging from the English tradition of political satire (including figures such as Gilray, Strube, Low and Rowlandson) to the anarchical farce of contemporary artists such as Steve Bell and Matt Wuerker (see Dodds 1996, 1998, O Tuathail, Dalby and Routledge 1998, Berg 2003). In recent years, political geographers have increasingly appreciated that cartoons can be considered in conjunction with other popular sources such as magazines, newspapers and television. This has also been matched with a growing interest in cultural geography for visual sources combined with a commitment to analyzing cartoons and images as 'texts' capable of multiple interpretations (see Crang 1998).

At first glance, the significance of humour and cartoons for international affairs may appear either marginal or trivial. It is quite common for scholars of world politics to produce weighty tomes on the condition of the interstate system and foreign policy without any illustrations or images. However, a number of authors have argued that political cartoons and images can be deployed as 'geopolitical texts', which illuminate or even subvert particular political practices such as foreign-policy decision-making. Political images can also be deployed to illustrate prevalent cultural anxieties about 'threatening neighbours' and 'dangers'. Critical geopolitical writers have also argued that cartoons can question and even transgress dominant relations of knowledge, truth and power (Berg 2003). British cartoonists such as Steve Bell deploy an 'anti-geopolitical' eye as they contest dominant representations of political affairs such as the 1982 Falklands War, the Bosnian crisis of 1992–5 and most recently the Bush administration's response to the September 11th attacks on the US (see Dodds 1996, 1998). Cartoonists often have their sketches positioned close to the editorials in daily newspapers and thus they help to shape the ways in which news and views are interpreted (Dougherty 2002: 258).

The term 'anti-geopolitical eye' was used by Gearoid O Tuathail to describe the critical writings and images of journalists such as Maggie O'Kane and cartoonists such as Steve Bell. Some of Bell's images are unsettling precisely because he uses satire, humour and shocking representations to question specific events. The best images and articles are acts of transgression in the sense that they call into question the dominant relations of power, knowledge and truth. Bell's images of Bosnia, for example, refused to accept the British government's assertion in 1993–4 that intervention on behalf of vulnerable civilians was politically unsafe and strategically unwise. His shocking portrayals of death and destruction helped to restore 'Bosnia' to our universe of obligation. As visual critiques, cartoons help to deconstruct the political agendas of political elites, national security bureaucracies and military officers.

In the midst of the Second World War, the *Daily Mirror*'s cartoonist Philip Zec published a famous cartoon of a ship-wrecked sailor clinging desparately to a piece of wood whilst floating in the middle of an ocean (see Fig. 4.8). The caption read 'The price of petrol has risen by a penny (Official)'. The cartoon coincided with a particularly traumatic moment in British politics when the *Daily Mirror* attacked the retention in Churchill's coalition government of ministers who had proposed to 'appease' Hitler in 1938. The prime minister, Winston Churchill, ordered a financial investigation of the *Daily Mirror*'s affairs in the light of criticisms of his ministers including Austin Chamberlain. The apparently unpatriotic behaviour of the newspaper was further condemned when Zec's cartoon was interpreted as being highly critical of the government's ability to deal with the German submarine attacks on British shipping in the North Atlantic. The home secretary, Herbert Morrison, came to Parliament and complained that, 'The cartoon in question is a particularly evil example of the policy and methods of a newspaper with a reckless indifference to the national interest' (1942, cited in Pilger 1998: 387). Morrison believed that the cartoon suggested that the lives of merchant seamen were being sacrificed in favour of the oil companies who would charge more for their products at a time of national shortage. Zec maintained that

Figure 4.8   Philip Zec's *Daily Mirror* cartoon of 1942
Source: Mirror Syndication International

he saw the image as highlighting the need to save petrol because importing goods and products cost lives. However, the *Daily Mirror*'s reputation for frank criticism meant that this cartoon became one of the most contested British images of the Second World War.

George Orwell's *Animal Farm*, published in 1945, was an allegory of the 1917 Soviet Revolution – a thinly disguised critique of the post-revolutionary Soviet Union and the corruptive practices of totalitarian governance. Orwell's hostility towards totalitarianism (regardless of whether it was on the political left or right) was shaped by his experiences of the Spanish Civil War in the 1930s when he fought with the republican forces (inspired by the ideals of democratic socialism) against the fascist forces of General Franco. The latter was to emerge as the authoritarian ruler of Spain until his death in the mid-1970s. Given the timing of its publication, the British Foreign Office recognized the potential of *Animal Farm* as a source of propaganda against the totalitarian Soviet Union. Officials in the International Research Department of the Foreign Office also commissioned Orwell to produce a cartoon-strip version of the book in order

Figure 4.9   A 1950s cartoon produced under the auspices of the Foreign Office's International Research Department
Source: The National Archives

that British embassies around the world could supply these images to local newspapers (Shaw 2001). It was noted in one Foreign Office memo dated from 1951 that, 'With a skillful story-teller one should have thought that it could be made into a very effective piece of propaganda down to *village audience level*' (my emphasis and cited in Norton-Taylor 1998: 7 and Fig. 4.9).

The phrase 'village audience level' becomes all the more pertinent if one considers what cartoons can achieve in a performative sense. Cartoons are a specific media form using humour and satire to convey messages about the social and political world (Dines 1995: 237, Dougherty 2002). The cartoon illustration of Orwell's *Animal Farm* shows a disaffected pig called Major dreaming of revolution ('Animals arise!') because of the apparent indifference of Farmer Jones to the welfare of his animals on the farm. British officials were convinced that Orwell's images were useful because they would appeal to the semi-literate village populations of strategically important countries such as India, Mexico and Brazil. Notwithstanding the patronising assumptions made about these Third World audiences, the warning of the cartoons and the book appeared to be profound: Western capitalist societies could not be complacent against the threat of Soviet totalitarianism, and totalitarianism was riddled with dangerous contradictions. Ironically, by the time of his premature death in 1950, Orwell had become disillusioned with British party politics and with the post-1945 government of Clement Attlee in particular.

The recently declassified secret papers of the British Foreign Office have revealed that George Orwell was not only responsible for naming other artists and writers as 'crypto-communists' but also for encouraging the then British government to engage fellow writers in a visual propaganda campaign against the Soviet Union. During the 1950s, a series of artists and writers were commissioned to produce cartoons and illustrations which sought to represent the Soviet Union as a dangerous and threatening place to Britain and her allies. Officials within the International Research Department of the Foreign Office also constructed other characters in order to illustrate the apparent dangers of communism and totalitarianism. One example was the invention of a character called Guy Greenhorn, who was depicted as a citizen of 'Democrita' – a 'likeable, intelligent but gullible young fellow and an admirer and sympathizer of communism'. He believed in the principles of 'Stalinovia' but was vulnerable to the influence of the evil-looking Dr Renegado. Greenhorn meets a young woman but eventually they become disenchanted with married life in Stalinovia. The tale concluded with the young Greenhorn and his wife returning to Democrita because they can no longer tolerate the restrictions of Stalinovia.

This must appear rather bizarre and even comical to readers at the start of the twenty-first century, but it should be remembered that British and American political life in the 1950s was dominated not only by the hope for post-war economic recovery but also by a fear of communism and the Soviet Union. The illustrated story of Greenhorn, which was released by the Churchill government in Britain, coincided with a series of high-profile defections to the Soviet Union by Cambridge University-educated Foreign Office officials such as Guy Burgess and Kim Philby. Amid near paranoia over intelligence leaks and communist penetration, cartoons and illustrated stories were seen as an important source for shaping the visual imaginations of citizens in the struggle against a Cold War adversary.

The spatial symbolics of humour can often challenge and even subvert dominant boundaries of national sovereignty and the nationalist scripting of place. During the 1982 Falklands War, for example, many editors felt obliged to adopt a position broadly supportive of Mrs Thatcher's decision to launch a task force to recover the Falkland Islands (see Fig. 4.10). The then government had invested considerable political and cultural capital in persuading editors and public opinion that whilst the islands were located 8,000 miles away from Britain they were in fact populated by loyal and patriotic British citizens. As Peter Jenkins of the *Guardian* noted in April 1982: 'By what weird calculus was it reckoned that the fate of all the free peoples might hinge upon the fate of 1,800 islanders and their 600,000 sheep?' (cited in Dodds 1996: 572). Alternatively, cartoonists such as Bell depicted the Falklands as a group of windswept rocks overwhelmingly populated by sheep and penguins. The presence of sheep, so familiar a part of the landscape for British readers, was an indirect means of ridiculing the British response to the loss of these islands at a time of high unemployment and social dislocation.

Geopolitics and music

As a form of human expression, music has been not only a powerful vehicle for the articulation of dissent and resistance (e.g. popular songs such as 'Free Nelson Mandela' and 'Sunday Bloody Sunday') but also a tool used by regimes and governments to gather support for particular forms (sometimes extremely violent and xenophobic) of nationalism (for instance, Nazi Germany in the 1930s used the stirring music of the German composer Richard Wagner to engender a sense of nationalism and patriotic loyalty). Popular musicians in America such as Bob Dylan were at the forefront

Figure 4.10 Steve Bell's 'If . . .' cartoon strip appeared in the centre-left British newspaper the *Guardian* during the 1982 Falklands conflict

Source: Copyright © Steve Bell 2000/All Rights Reserved

of anti-Vietnam protests in the 1960s. More recently, music and musicians have been at the centre of controversy, as witnessed during the 1991 Gulf War when British DJs were instructed not to play songs such as Abba's 'Waterloo' for fear of causing offence to the Allied troops stationed in Kuwait and/or their families. The lyrics of the song were considered not only excessively militaristic but also potentially insensitive to French troops serving in the Gulf because 'Waterloo' referred to an Anglo-German victory over the French in 1815. At the same time, musicians in Algeria and Afghanistan were murdered because the extremist authorities believed that music was a form of Western pollution, which bred dissent and disrespect for Islam and the Islamic way of life.

In recent years, Anglo-American geographers have begun to explore how music contributes to specific constructions of place and cultural identity (see Smith 1994, Leyshon, Matless and Revill 1995, Connell and Gibson 2003). In particular, this geographical research has emphasized how music (in all forms from classical to rave) has been neglected in the social sciences and humanities in favour of more visual sources such as film, cartoons, television, paintings and landscapes (Smith 1994: 235). In the last ten years, however, geographers have produced some interesting research which seeks to locate music in its geographical, political and cultural contexts (see In focus 4.6). In nineteenth-century Britain, for example, the brass-band movement forged and sustained community music-making in industrial areas dependent upon

---

In focus 4.6: Nitin Sawhney and *Beyond Skin* (1999)

The British-Asian artist Nitin Sawhney provides an interesting example of how music can be used to reflect on identity politics and in this case the nuclear nationalism of India in 1998. His album *Beyond Skin* starts with the words of Prime Minister Vajpayee of India announcing the testing of three nuclear bombs in May 1998. The prime minister urged all Indians to celebrate this technological/military achievement even if it led to Pakistan launching its own nuclear testing in retaliation and to widespread international condemnation. Was it unpatriotic to be critical of the testing? Forty years earlier, Robert Oppenheimer, the leader of the US Manhattan Project, quoted the sacred Hindu text the *Bhagavadgita* in the aftermath of the first nuclear explosion in July 1945 – 'Now I am become death, the destroyer of worlds'. As Sawhney's lyrics reflect, an American scientist quoted a Hindu text to condemn the Bomb and an Indian prime minister (and Hindu nationalist) celebrated nuclear weapons as an expression of 'Western' progress.

But as Sawhney notes, the relationship between the West and India also has implications for his own sense of personal identity. He lives in a country (the UK) which defines him by the colour of his skin (Asian), and hails from a country which would define him by his religious heritage (Hindu). As he states on the insider cover of the album, 'My identity and my history are defined by myself – beyond politics, beyond nationality, beyond religion and *Beyond Skin*.'

extractive industries such as coal, iron and steel. Amid mass unemployment and social dislocation in the late twentieth century in the aftermath of industrial decline, the brass-band movement helped to consolidate communal loyalties and a collective sense of purpose (see S. Smith 1994, 1997).

Human-rights organizations consider music to be one of the most censored forms of art. During the Cold War, for instance, dissident bands in communist Eastern Europe (such as Plastic People of the Universe in Czechoslovakia) released their music via underground recording labels in order to protest about all forms of censorship and human-rights violations. Popular music was considered especially subversive because of its appeal to young people. The Czech president, a former imprisoned dissident poet, Vaclav Havel, revealed in 1989 how he had been inspired by the music of Lou Reed. In South Africa, the American musician Paul Simon joined forces with the black band Ladysmith Black Mambazo to protest through music at the continued injustices and inequalities of apartheid. During the repressive era of General Pinochet's Chile (1973–89), the security forces killed the folk singer Victor Jara in a Santiago football stadium in 1975 because he was considered a left-wing subversive.

Thus far, geopolitical writers have not fully explored the geographical soundscapes and political worlds created by music. On a regional scale, research into the voting patterns for the Eurovision Song Contest (created in 1956) suggests particular cultural and geopolitical factors (Yair 1995). In 2003 the British group Gemini received no votes and some commentators suggested that this was partly due to European voters wishing to register their disapproval of the UK's support for the invasion of Iraq. In other words, a British music entry was being 'punished' by association. Others have noted that there is a longer history of hostile and or supportive voting depending on the countries involved. For example, Greek and Turkish voters do not tend to favour one another's entries because of their long-standing enmities over the divided island of Cyprus, whereas German voters tend to be highly supportive to the Turkish entries (because of a large Turkish community living in Germany).

A more specific case for consideration would be the music of the Irish band U2, created and sustained by a particular representation of British violence in Northern Ireland. Their best-selling album *Under a Blood Red Sky* (1981) opens with the now immortal line 'This is not a rebel song, this is Sunday Bloody Sunday'. It is ironic because the song ('Sunday Bloody Sunday') refers to the shooting of 13 unarmed nationalist (Catholic) civilians by British paratroopers in January 1972 (called Bloody Sunday by the media) in Northern Ireland. For many nationalists this massacre further cemented the view that Northern Ireland was a province controlled by a protestant/unionist majority using the armed forces and police service to oppress and even murder a religious minority. For Irish people (from the North and South of the island of Ireland), rebel songs (many originating from the bloody civil wars of the 1920s) have served to express dissent and resistance against the continuing British colonization and occupation of Northern Ireland/Six Counties/Ulster.

In another instance of politicized music, punk bands in Germany generated a following of sorts amongst young (white) Germans attracted to lyrics blaming unemployment and poverty on immigrants and foreigners. While it would be wrong to imply that this music directly contributed to outbreaks of violence against Turkish families in Germany, there is no doubt that music can evoke powerful feelings and provoke movements and acts of political resistance. Music can help to create particular political and symbolic geographies of resistance, censorship and expressions but it can also be a matter of life and death in countries such as Algeria, Burma and China. The death of the singer Boudjema Bechiri in Algeria in 1996 was one example of a musician being murdered by a brutal government seeking to suppress any form of dissent or resistance in the midst of Algeria's bloody and ongoing civil war (see Chapter 8).

Conclusions

Popular geopolitics is an emerging field of interest within critical geopolitics. Drawing on media theory and related disciplines such as International Relations, it seeks to explore how the media contribute to the representation and interpretation of global political space and associated events. It recognizes that the media including newspapers on the one hand can contribute to the projection and reinforcement of particular national and or transnational identities and ideologies, and yet on the other hand, help subvert and contest such hegemonic positions (Fig. 4.11).

Figure 4.11   The Kenyan national newspaper *The Nation* has been an important source of information for Kenyans and a medium through which governments have sought to forge national unity
Photo: Klaus Dodds

This chapter has illustrated how various media can be used to examine how different forms of communication and imagery represent the social and political world. The potential effectiveness of films can be gauged through the varied geographies of dissemination and distribution, which ensure that movies such as *Top Gun* (1986) and the James Bond films are seen not only in the cinema but also via commercial television and video rental. The role of the televisual media in representing and portraying international events ranging from war, humanitarian disasters, international conferences to (televisual) diplomacy has become commonly accepted as pundits and observers recognize the power of television pictures and soundbites within world politics. In a similiar vein, magazines and cartoons equally contribute to the communication of ideas and interpretations of world affairs and specific places and peoples. The focus on visual spectacles such as war and disasters may mean, however, that other kinds of events such as routine economic processes are neglected as a consequence.

Key questions

- Where do most people in the North learn about 'foreign' news? Is that different to people living in the Global South?
- Why do popular representations of danger matter?
- What kind of dangers and threats have replaced the Soviet Union in post-Cold War American cinema?
- Do the mass media when reporting events such as the US-led assault on Iraq in 2003 help to 'manufacture consent' in liberal democracies?

Further reading

There is a massive amount of literature available on the various forms of media discussed. For film and television in general see P. Harrison and R. Palmer, *News Out of Africa* (London, Hilary Shipman, 1986) and C. Barker, *Global Television* (London, Routledge, 1998). J. Der Derian, *Virtuous War* (Boulder, Westview, 2001) is highly significant in tracing the connections between Hollywood and the US military. On representation and the cinema see E. Shohat and R. Stam, *Unthinking Eurocentrism* (London: Routledge, 1994), D. Morley and K. Robins, *Spaces of Identity* (London, Routledge, 1995), K. Robins, *Into the Image* (London, Routledge, 1996) and P. Davies and P. Wells (eds.), *American Film and Politics from Reagan to Bush Jr.* (Manchester, Manchester University Press, 2002). On the importance of cartoons see B. Dougherty, 'Comic relief: using political cartoons in the classroom', *International Studies Perspectives* 3 (1995): 258–70. An excellent collection of cartoons and commentary on the Anglo-Irish question can be read in R. Douglas, L. Harte and J. O'Hara (eds.), *Drawing Conclusions: A Cartoon History of Anglo-Irish Relations 1798–1998* (Belfast, Blackstaff Press, 1998). On the analysis of music, see A. Leyshon, D. Matless and G. Revill, 'The place of music', *Transactions of the Institute of British Geographers* 20 (1995): 423–33. On the geopolitical significance of popular sources see G. O Tuathail, S. Dalby and P. Routledge (eds.), *The Geopolitics Reader* (London, Routledge, 1998), M. Crang, *Cultural Geography* (London, Routledge 1998). J. Sharp, *Condensing the Cold*

War (Minneapolis, University of Minnesota Press, 2000) provides an excellent discussion of the *Reader's Digest* magazine. For more information on the Internet see B. Wellman and C. Haythornthwaite (eds.), *The Internet in Everyday Life* (Oxford, Blackwell, 2002).

Websites

Al Jazeera www.al-jazeera.com
CNN International www.cnn.com
James Bond www.jamesbond.com
Internet Movie Data Base www.imdb.com

Chapter 5

GLOBALIZATION OF DANGER

Key issues

- Why does the ownership of weapons of mass destruction (WMD) carry geopolitical implications?
- How do states and non-state organizations try to control weapons proliferation and nuclear testing?
- Why did some Third World states and NGOs try to resist nuclear testing?
- What role does regional competition between states play in encouraging WMD proliferation?

Consider the following statement made by a senior member of the Bush administration in July 2002. US Secretary of State Donald Rumsfeld testified to the Senate Foreign Relations Committee that:

> Some have asked why, in the post Cold War world, we need to maintain as many as 1700–2200 operationally deployed warheads? The end of the Soviet threat does not mean we no longer need nuclear weapons. To the contrary, the US nuclear arsenal remains an important part of our deterrence strategy, and helps us dissuade the emergence of potential or would-be peer competitors, by underscoring the futility of trying to reach parity with us (cited in Shambroom 2003: ix).

Despite spending no less than $5 trillion (since 1945) in maintaining an extraordinary armory against the possible threat of a Soviet attack, the US still perceives a strategic need for Cold War levels of preparedness. According to this line of geopolitical reasoning, the US must maintain constant vigilance against new threats, whether in the form of rival states (not specifically enumerated) or terror groups such as the Al-Qaeda network. Billions of dollars will be devoted to a new National Missile Defense (NMD) scheme, as concern is expressed over the possible antics of 'nuclear outlaw' states such as North Korea and Iran.

The tragic September 11th attacks on the United States proved that those intent on causing harm do not need to use nuclear weapons. Two planes (with the combined impact of small tactical nuclear weapons) destroyed one of the most enduring symbols of urban New York. Later that month, five people died as a result of unrelated anthrax attacks in the Washington DC region.

The threat posed by nuclear weapons and other WMD such as chemical and biological weapons continues to be enormous even if geographically varied. The United States deployed 18 million gallons of Agent Orange to poison the ecosystems of Vietnam in the 1970s, and Saddam Hussein's regime in Iraq attacked Kurds and Iranians with chemical weapons in the 1980s. Even apparently isolated experimental nuclear explosions in the South Pacific have destroyed island ecosystems. The deadly power of nuclear bombs gave rise to a host of anti-nuclear movements in North America and Western Europe which argued that Cold War ideologies of nuclear deterrence contributed to further insecurity in the world with the creation of a nuclear culture which dehumanized opponents, exaggerated threats to national security and downplayed the consequences for human and environmental life (Lifton and Falk 1982, Beck 1992). The effects of test programmes on people living in the South Pacific, the centre of Australia and the allegedly remote parts of the USA and the former Soviet Union have only been formally documented in the last ten to fifteen years. For many people affected by nuclear testing, nuclear war was not unthinkable.

Geopolitics and weapons of mass destruction

The destructive powers of nuclear weapons/WMD remain a graphic symbol of a fragile planet potentially threatened by annihilation. Within a geopolitical framework, the spread of nuclear, chemical and biological weapons and the geographies of nuclear testing have a powerful impact on global politics.

1. For realists and traditional geopolitical writers, WMD proliferation demonstrates how powerful states have sought to develop WMD diplomacy in order to protect their national-security interests (see Waltz 1981). The major nuclear states include the USA, Russia, France, Britain, China, India, Pakistan, North Korea and Israel.
2. For the liberal institutionalists, the capacity of the international community and organizations such as the United Nations to deal with nuclear weapons has been stretched. Overall, however, the UN has managed to create a negotiating atmosphere conducive to a measure of control and disarmament even if plans to develop a NMD by the United States threaten this potential (see Keohane and Nye 1989).
3. For feminist observers, both realist and liberal institutionalist analyses of nuclear weapons and proliferation neglect how the experiences and consequences of these issues vary between men and women and from place to place. It has also been noted how the geographies of nuclear testing were felt most strongly in the territories of tribal peoples and ethnic minorities in the USA, Australia, China and the South Pacific (see Enloe 1989, 1993).
4. Critical geopolitical writers have highlighted how nuclear weapons raised important questions about the nature of global political life and the unequal geographies of nuclear threats and testing (see Dalby 1990). They have also questioned the objectivity of powerful WMD states such as the US when it

comes to labelling other nuclear weapons states such as North Korea as 'rogue states' or 'outlaw' states and thus justifying either aggressive action against those states or WMD proliferation.

In the post-Cold War era, new hopes emerged that nuclear weapons could be abolished if the superpowers agreed to major cuts in their stockpiles and to the removal of weapons systems from Western and Central Europe. Sceptics argue that nuclear proliferation still remains a serious problem for global politics, as witnessed by the ongoing attempts of Israel, North Korea, India and Pakistan to consolidate their nuclear capabilities and the discovery of Iraq's clandestine nuclear weapons facilities in the aftermath of Operation Desert Storm (Gardner 1994, Mazarr 1995, Cirincione 2000). On the one hand, realists claim that nuclear weapons cannot be removed from the international political scene because they cannot be 'un-invented', but on the other hand, a number of critics insist that this sort of pessimism could be overcome by developing rigorous global disarmament regimes and by generating a global consensus to denounce nuclear and other weapons of mass destruction. In the midst of US paranoia about terror groups and 'rogue states', critics have posed the following questions: Why would a small state such as North Korea risk obliteration by attacking the United States? If the Soviet Union respected the logic of deterrence (i.e. by not attacking the USA because of the terrible consequences for itself) then why would the same not be true for small nuclear-weapons states? Why would terror groups need WMD when Al-Qaeda used conventional weapons in the form of fuel-heavy aeroplanes to devastating effect?

Global politics, nuclear weapons and the nuclear weapons cycle

Since 1945 the international community has faced the possibility of global nuclear war. The exchange of ideas and people between Nazi Germany and the United States provided a vital link in the development of nuclear weapons. The major figures behind the top-secret US Manhattan Project and the first nuclear explosion in New Mexico in July 1945 were German-speaking scientists who had either fled their native European countries during the Second World War or were recruited by the Americans at the end of the conflict. Despite claims that nuclear technology would benefit world civilization more generally, the United States had ensured that the twentieth century would be immortalized as a dangerous epoch. The American bombing of Nagasaki (the Fat Man bomb) and Hiroshima (the Little Boy bomb) in August 1945 demonstrated the deadly potential of nuclear explosions even if it was seen as a justifiable action which sought to bring the Second World War to an end (see In focus 5.1).

It is estimated that in terms of immediate deaths, 70,000 died in Hiroshima and 35,000 perished in Nagasaki (see Fig. 5.1). The overall death toll is unknown because many of the survivors later died from ill health relating to nuclear radiation.

In 1946 the US government recruited German military scientists to assist in the development of nuclear technology, rocket and space technology. Within three years of

Figure 5.1   The defining image of the twentieth century? the nuclear bomb
Photo: © Rex Features

In focus 5.1: Ground Zero, Nagasaki

A US navy officer stationed in Nagasaki wrote the following to his wife in September 1945:

> A smell of death and corruption pervades the place, ranging from the ordinary carrion smell to somewhat subtler stenches with strong overtones of ammonia (decomposing nitrogenous matter, I suppose). The general impression, which transcends those derived from the evidence of our physical senses, is one of deadness, the absolute essence of death in the sense of finality without hope of resurrection. And all this is not localized. It's everywhere, and nothing has escaped its touch. In most ruined cities you can bury the dead, clean up the rubble, rebuild the houses and have a living city again. One feels that is not so here. Like the ancient Sodom and Gomorrah, its site has been sown with salt and 'icabod' [the glory has departed] is written over its gates.

Source: Richard Rhodes 'Introduction' in Shambroom 2003

---

the American nuclear bombing, the newly created United Nations Commission for Conventional Armaments decreed a new category of weapon: weapon of mass destruction (WMD). By 1964, five countries had tested either fission or fusion nuclear weapons: USA, USSR, UK, China and France. The *nuclear weapons cycle* (NWC) refers to the production, testing and deployment of nuclear weapons. While it is common to focus on the production and deployment of such arms systems, the geographies of testing and their social and cultural implications must not be neglected in this analysis of world politics. Nuclear-weapons testing occurred in many parts of the world, often in places perceived to be marginal in the minds and actions of national elites. As is still the case today, issues around the development of nuclear weapons sometimes became entangled with those concerning nuclear energy. For the first decade of the so-called post-war period many heads of state viewed the invention of nuclear energy as a positive sign: a source of endless and cheap energy and a symbol of the modern technological state. For example, President Truman's 'atoms for peace' speech in 1953 was similar in tone to his earlier speech in 1949 extolling the virtues of modern industrial-based development (see Chapter 3). Unfortunately, his speech coincided with considerable tension as the rival superpower, the Soviet Union, carried out high-profile nuclear testing in Siberia. Concerns about the Soviet nuclear arsenal notwithstanding, successive American governments remained committed to developing a civil nuclear energy programme, and this was to become steadily more significant in the 1960s and 1970s as fears grew about excessive dependency on imported oil.

Global and national institutions have struggled to cope with the legacy posed by nuclear weapons through either arms control or disarmament. How do you control such a process in the absence of a global authority with the power to regulate the activities of powerful states such as the United States and the Soviet Union? Attempts to regulate have focused on international regime-building and diplomatic intervention involving the NWS. Arms control is a process of gradually limiting and/or restraining production, testing and deployment of weapons, while disarmament is a more radical proposition concerned with the renunciation of weapon systems. In realist thought, nuclear weapons and nuclear culture have been endowed with qualities which emphasized the enhanced military and diplomatic capabilities afforded to a state within the anarchical international arena.

In that sense, the acquisition of nuclear-weapons technology by the UK in the late 1940s must be considered indicative of a desire by the Churchill and Attlee governments to remain a major power on the world stage in the aftermath of the Second World War. With the emergence of the two superpowers, many British public figures were arguing that nuclear weapons would improve Britain's standing in the world, especially in the wake of anti-colonial violence and the loss of imperial possessions such as India in 1947. By the mid-1960s, all five permanent members of the UN Security Council (China, France, the Soviet Union, the UK and the USA) had obtained the status of nuclear-weapons state (NWS). The possession of nuclear weapons was seen to enhance national prestige and political influence, even if very few NWS ever thought that they would launch an attack on another nuclear power.

Realists argue that the possession of nuclear weapons offers certain advantages in terms of bargaining power and international profile. However, there is also a strong sense that the pursuit of national interest and national survival would not be facilitated by an aggressive deployment of nuclear weapons. Deterrence is more sensible than active deployment, given the capacity of nuclear weapons to cause massive damage. In the words of some writers, a form of 'nuclear taboo' has existed during the last fifty years in which states have sought to engage in confidence-building measures and arms-control processes in order to lessen the dangers of a potential nuclear war. It was argued, therefore, that nuclear weapons provided a form of stability for the international system during the Cold War period. Hence, there has been widespread concern amongst members of the George W. Bush administration (2000–) that the spread of NWS will actually destabilize the 'nuclear club' and global deterrence.

In conjunction with liberals, realists also believe that the arms-control process sponsored by institutions such as the United Nations and specialist international agencies can help to mediate the dangers posed by nuclear weapons. In the 1960s, for example, the world stood on the brink of a nuclear confrontation on three separate occasions (see In focus 5.2).

The United Nations charter commits states to pursuing peace and common security under Article 11 and the UN has sought to advocate solutions to specific problems and to sponsor resolutions on nuclear testing, deployment and the peaceful use of nuclear energy. Both realists and liberal institutionalists endorse the significance of confidence-building as well as international diplomacy and designated nuclear conferences to generate a consensus on such issues.

For the globalization theorist, nuclear weapons provide evidence of the spread of technology (horizontal proliferation) and danger across the planet. Having rejected

---

In focus 5.2: Near nuclear confrontations in the 1960s

1. August 1961 **The Berlin Crisis**. Soviet and East German forces began to construct a wall that divided the city of Berlin into Western and Eastern sectors. Tension mounted over whether the Western Allies (UK, US and France) would be able to maintain access to the city in the light of attempts by the Soviets to restrict access.
2. October 1962 **The Cuban Missile Crisis**. Soviet attempts to establish a missile station on the island of Cuba provoked the Kennedy administration to blockade Soviet transporters from delivering the final elements for the construction of the station. Unbeknown to Kennedy, the Soviets already had a limited nuclear capability on Cuba. After a tense stand-off, the Soviets ordered the ships to turn around and return to the Soviet Union and thus averted a possible American attack on Cuba. America later withdrew its Jupiter missiles from Turkey in an unofficial truce.
3. June 1967 **The Six Day War**. US-backed Israeli and Soviet-backed Arab forces became embroiled in a third war since the creation of Israel in 1948. Within a week the superior Israeli forces routed their opponents and occupied the Golan Heights in Syria, the Sinai Desert and the West Bank. The Soviets warned against further military conquests in the light of their concerns about the strategically sensitive Middle East.

the overwhelming focus on political power and the state, globalization and feminist theorists on world politics emphasized the complexity surrounding nuclear issues. Social and political life, the state and international relations have all been profoundly affected by the intensity of the nuclear weapons cycle. In the post-war period, the NWS began a cycle of extensive nuclear testing in the Third World, remote parts of China and the Soviet Union, or places populated by indigenous peoples. In the period between 1945 and 1991, over 1,900 nuclear explosions were carried out on the earth's surface. The leading testers were the USA (936), the Soviet Union (715), France (192), UK (44) and China (36). The primary sites of these tests were Kazakhstan (467), French Polynesia (167), the Marshall Islands (66), Xinjian province of China (36) and other sites such as indigenous peoples' reservations in Nevada and Australia. The effects of nuclear testing are still being felt in the communities of many of these islands, remote provinces and interior reserves (Kato 1993).

The impacts of nuclear testing on the affected communities have been well documented in the case of the South Pacific (see Fig. 5.2). Joni Seager noted, 'Most of the nuclear weapon systems stationed in Europe and the US have been tested on "indigenous peoples" lands in the Pacific, without their consent and often without warning' (Seager 1993: 61). In the 1950s, for example, Britain carried out nuclear testing in the centre of Australia and on Christmas Island in the Pacific Ocean. With the assistance of the Australian government, 800,000 square kilometres of Aboriginal land in Northern Territory was given over to nuclear testing without any form of compensation for those who were effectively dispossessed. In other parts of Australia such as Pine Gap and Nurrungar, the Americans created top-secret surveillance bases, which were off limits to all Australian citizens. Most of the nuclear test sites remain highly contaminated

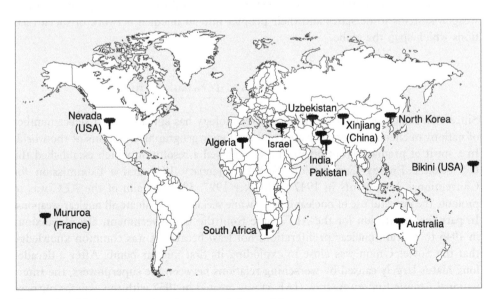

Figure 5.2   Major nuclear test sites around the world

and both British and Australian governments are reluctant to address the substantial issues of cleaning up the sites and compensating the victims.

By a further twist of irony (see below), the decision to declare Antarctica a zone of peace and co-operation in 1959 was only made possible because the three major powers had secured alternative sites (often inhabited by indigenous populations) such as Nevada, Siberia, the Pacific Ocean and the Australian continent for test purposes. Five years earlier, in 1954, the American armed forces had bombed the Bikini Islands using 17-megaton hydrogen bombs in order to assess the likely impact of a bomb 1,000 times more powerful than those dropped on Japan in 1945. The islands have now been declared uninhabitable for 30,000 years (Seager 1993: 64) and the original residents have been relocated to various other island groups without compensation.

Most standard realist texts on nuclear strategy and weapons have little time or scope for the local consequences of nuclear testing. The combined effects of British, French and American testing in the South Pacific are a footnote in the accounts of weapon systems and their effect on the international scene. The key dimensions of inter-national life remain military force, war, the state and the anarchical international arena. Feminist and anti-nuclear observers would argue that women and ethnic-minority groups have often borne the brunt of this vision of political life (Pettman 1996). The litany of cancers, deformed babies and premature death in these areas is truly shocking. Govern-ments have for far too long refused to accept responsibility for their actions. At the time of writing, the Australian, UK and US governments are negotiating compensa-tion packages with groups of indigenous peoples affected by nuclear testing. While liberals have tended to emphasize the capacity and willingness of the international community to engage in negotiation and confidence-building measures in an attempt to prevent nuclear weapons from becoming an accepted and widespread feature of global political life, critics point to nuclear-testing stories to illustrate the capacity of these weapon systems to tie states and their peoples into an unequal network of social rela-tions which span the globe.

Post-1945 nuclear proliferation control

Since 1945, nuclear and ballistic missile technology has spread from a select number of nations to encompass and/or affect all the major geographical regions of the world. In a spirit of premature optimism, the UN passed a resolution which established the UN Atomic Energy Commission (AEC) in January 1946 and a Commission for Conventional Armaments in 1947 (Whitaker 1997: 58). The aim of the AEC was to promote the peaceful use of nuclear energy while seeking to eliminate all nuclear weapons. In part the inspiration for the AEC came from the US government, already anxious in 1946 to restrain nuclear proliferation, not least because it was common knowledge that the Soviet Union was close to exploding its first nuclear bomb. After a decade-long hiatus largely caused by worsening relations between the superpowers, the Inter-national Atomic Energy Agency (IAEA) was created in 1957 with the express purpose of monitoring nuclear-testing programmes. In the midst of the Cold War, however, the

IAEA was beset with difficulties because the NWS were wary of allowing international inspectors access to their secret nuclear plants and storage depots.

In spite of the difficulties over monitoring, the nuclear powers agreed to a voluntary moratorium on testing between 1958 and 1961 and also secured the status of Antarctica as the world's first nuclear-free zone. Under Article V of the 1959 Antarctic Treaty, all the signatories (including the USA, the UK, the USSR and France) agreed to prohibit all forms of nuclear explosion, test and waste disposal. This achievement needs to be tempered with the realization that Antarctica's status as a non-nuclear test site was secured at the expense of many other inhabited places in the world. The IAEA had the difficult task of monitoring testing programmes whilst at the same time introducing systems of audit, report and inspection amongst the NWS. The Vienna-based agency was commissioned to promote the peaceful application of nuclear energy at a time when the superpowers were busy stockpiling weapons and expanding the geographical deployment of these systems.

In spite of these pressures, the international community achieved a measure of progress on arms control through a series of treaties in the 1960s and 1970s.

1963   Limited Test Ban Treaty signed between the US and the USSR, which banned testing in the atmosphere, underwater and in space but permitted underground testing.
1964   Treaty of Tlatelolco (Mexico) signed by 23 Latin American states in order to promote regional non-proliferation.
1968   Nuclear Non-Proliferation Treaty signed by over 100 states including the superpowers but excluding NWS such as France and China. The treaty was ratified in 1970.
1976   Outer Space declared a Nuclear Weapons Free Zone (NWFZ), treaty signed by 113 states.

The NNPT Treaty of 1968 was critical because it provided a benchmark for nuclear proliferation control and strengthened the role of the IAEA in terms of preventing the illicit use of nuclear technology. Agreement was reached that a review conference on nuclear use would be held every five years for the purpose of investigating and assessing the state of nuclear knowledge and the adoption of peaceful uses of nuclear energy.

The nuclear non-proliferation and disarmament process was undoubtedly assisted by a number of agreements, which included:

1971   Indian Ocean declared a zone of peace.
1972   Nuclear testing on the sea bed banned within a 12-nautical-mile zone of territorial waters.
1974   Threshold Test Ban Treaty between the two superpowers which limited testing to a 150 kilotons explosive yield.
1985   South Pacific declared a zone of peace.
1986   Intermediate Nuclear Forces Treaty between USA and USSR.
1997   Comprehensive Test Ban Treaty. All forms of nuclear testing banned.

Attempts to control the spread of nuclear weapons were hampered by the contested politics of the Cold War, when with the help of a number of Western European suppliers like France and Germany, 'nuclear threshold' states such as South Africa, Israel and Argentina developed nuclear programmes. Whereas the USA was anxious not to supply nuclear technologies to the Middle East and Latin America, German and French companies assisted in the construction and production process. It is alleged, for example, that this enabled the South African and Israeli governments to produce a large and unexplained nuclear explosion in the Southern Ocean in 1979. Two years later, Israel launched an illegal bombing raid on Iraq's German-made nuclear complex in Bushehr because of fears that an Iraqi nuclear programme would compromise Israeli national security. While this completely violated the norms of international law, it was justified by reference to national self-defence. Key allies such as the United States refused to condemn Israel in the United Nations. At the same time, NWS issued a number of so-called negative security assurances that no nuclear weapons would be used against non-nuclear weapons states. Only China, however, declared at the 1978 UN Session on Disarmament that it would never use nuclear weapons (as first-strike weapons) on any state unless it was attacked by a NWS.

Over the last 20 years, the NNPT has been widely recognized as the major mechanism for global nuclear disarmament. The 1968 Treaty stipulated that after a period of 25 years, participants were entitled to call for a review of the NNPT. The fifth major review conference (held in 1995, in New York) was critical because 185 members voted to extend the treaty indefinitely. On a less positive note, however, calls for a Middle Eastern nuclear-free zone were not endorsed by the treaty parties in spite of strong support from other countries in regional nuclear-free zones such as Latin America and the South Atlantic. The main sticking points were the refusal of Israel to dismantle her unconfirmed nuclear capability until regional peace was secured and Iraq's continuation with WMD programmes in violation of UN Resolution 1441, which demanded that Iraq allow UN inspection teams to verify disarmament.

Notwithstanding the commitment of the NNPT to non-proliferation and disarmament, a number of outstanding issues remain. The first issue concerns compliance and refers to states which have already 'given up' their nuclear weapons such as Ukraine and South Africa and to those who refused to accept the authority of the IAEA such as North Korea. In 1993 President de Klerk announced that South Africa had constructed six nuclear devices, all of which would be dismantled, and this was confirmed under the offices of the Mandela government. Other states such as North Korea and Iraq (until 2003) continued to cause problems for international inspectors because of their refusal to allow unfettered routine examinations of their nuclear operations. Israel, a widely suspected NWS, has avoided inspection by the IAEA. The second major concern regards the relationship between the established NWS and the new NWS such as India and Pakistan, which maintain that the USA has been hypocritical in seeking to restrict the nuclear ambitions of South Asian and Middle Eastern states while condoning those of Israel. Unless and until the world commits itself to complete nuclear disarmament, Indian and other political figures remain reluctant to end their weapons development programmes. Some realist writers such as Kenneth Waltz have even suggested that the further proliferation of nuclear weapons might actually stabilize world politics if the capacity for such weapons is spread more widely in the international arena.

The nuclear non-proliferation process is intimately linked to nuclear testing. The NNPT conference in New York recognized that advocates in favour of eventual nuclear disarmament had to recognize that there was a wider nuclear weapon cycle. The major problem concerning nuclear testing is that treaties signed in the 1970s, such as the Threshold Test Ban Treaty and the Peaceful Nuclear Explosions Treaty, have not been enforced in spite of appeals by the UN Disarmament Commission and the Conference on Disarmament. Both these treaties attempted to restrain test explosions and sought to implement a system of verification amongst the international community and interested observers. It was only in the aftermath of the Cold War that further progress was possible on nuclear testing. Notwithstanding fears of confrontation in South Asia, major dilemmas for the IAEA and the CTB parties remain, such as who is best suited to oversee the ban on all forms of nuclear testing and what sort of controls/constraints could be imposed by the international community on states which breach the testing ban. Ultimately, the biggest question of all is this: will the established NWS give up their nuclear weapons and commit themselves to permanent disarmament?

Resisting nuclearization: regional initiatives and nuclear criticism

In 1959, 12 nations signed a landmark treaty in Washington DC which created a legal and political framework for the management of the Antarctic continent and surrounding seas. The major elements of the Antarctic Treaty included: establishing scientific activity as the major concern of interested parties, preventing all forms of nuclear testing, dumping and explosions in the region and creating a forum where the environmental protection of Antarctica would be a priority. Previous territorial claims to the continent were considered suspended for the duration of the treaty. Since its ratification in 1961, 43 member states have accepted the principles of the Antarctic Treaty and subsequent additional measures and protocols. Western powers celebrate the 1959 Antarctic Treaty for declaring the Antarctic a NWFZ, and while this was a laudable achievement it did nothing to protect territories such as South Pacific Islands and the Siberian interior where tests continued to be conducted.

Carol Cohn's ethnographic analysis of strategic culture in the US demonstrated that the language of techno-strategic discourse tended to underplay the dangers of nuclear weapons and generated abstractions and euphemisms (such as calling nuclear bombs 'Fat Man' and 'Little Boy') which failed to connect the deadly potential of bombs to everyday life. During the 1950s, this sense of unreality became evident when scores of American school children were instructed about civic defence in the nuclear era (see Cohn 1987). On being told that a nuclear attack was imminent, children were expected either to hide under their desks or if they were playing outside to lie down immediately.

Helping to perfect this routine was a cartoon character called Bert the Turtle who advised children to DUCK and COVER! Bert seemed to imply that if children averted their gaze from the bright flash associated with an exploding nuclear bomb they were likely to survive. By the early 1960s, Bert was in retirement.

In contrast, a form of extra-national discourse based on anti-nuclearism has been a significant feature of the English-speaking world since the late 1950s. Peace movements such as the Campaign for Nuclear Disarmament (CND) have stressed the global dangers of nuclear weapons and poured scorn on the idea that the Bomb has contributed to greater global security (see Seager 1993, Pettman 1996). Public demonstrations and marches in the UK and the USA added to a body of opinion in these societies that nuclear weapons were dangerous and directly contributed to further global tensions. Popular culture was also beginning to reflect the very real dangers posed by a world made mad by the extremes of nuclear weapons (see In focus 5.3).

In his book *The Control of the Arms Race*, the British scholar Hedley Bull noted that techno-strategic discourses on nuclear weapons tended to assume that security referred to the state of superpower relations rather than the wider world (Bull 1965). However, by the 1960s it was abundantly clear that the possibility of nuclear war threatened all citizens regardless of their location. The development of rocket technology and

---

In focus 5.3: Atomic cinema

Produced by Stanley Kubrick, the film *Dr Strangelove: Or How I Learned to Stop Worrying and Love the Bomb* (1964) is probably the most memorable representation of nuclear paranoia. The actor Peter Sellers played three roles: a mad US general, a British air force officer and Dr Strangelove, a crippled former Nazi officer who cannot prevent involuntary Heil Hitler salutes. The plot revolved around the US general ordering an attack on the Soviet Union in response to fears that the Soviets have secretly poisoned him. In a desperate attempt to foil the attack the US president begs his Soviet counterpart to help shoot down the planes and thus prevent the nuclear destruction of the Soviet Union. Unbeknown to the mad general, the Soviets possess a 'Doomsday Machine' which will destroy the world once the Soviet Union is attacked. One US plane manages to make it through to the Soviet Union, and the film ends with the destruction of the world. In the background, the famous Second World War entertainer Vera Lynn can be heard singing 'We'll meet again'.

The film brilliantly exposed the paranoia and madness associated with the nuclear era, not least because it was filmed in the claustrophobic environment of an air force base, the war room and a bomber. It not only explicitly mocked American propaganda involving 'Duck and Cover' civil defence but also highlighted the uncontrollable dangers posed by the military-industrial complexes of the superpowers.

Released in January 1964, *Dr Strangelove* proved highly popular in an era dominated by the near nuclear confrontation in Cuba (1962) and amidst widespread fears in the US that a 'missile gap' was developing its arsenal and that of the Soviet Union.

Further reading: *see* Newkey-Burden 2003: 45–6.

nuclear delivery systems facilitated global access as nuclear missiles could, theoretically, travel across national boundaries. Talk of security in terms of superpower relations marginalized the condition and prospects of the vast body of humanity. It also revealed the profound inequalities between nations, as some states such as the USA enjoyed considerably more advantages not only in terms of nuclear stockpiles but also in the deployment of surveillance technologies.

Criticism by peace campaigners and academics began to gather momentum in the 1980s. Representatives of CND began demonstrations and a long vigil outside the American airbase at Greenham Common in Berkshire after Prime Minister Thatcher's decision to allow the deployment of Cruise missiles in the UK as part of the global defence of the 'Free World'. Women picketing military sites such as Greenham Common in the UK and Pine Gap in Australia were often labelled 'hysterical' and 'mad' by the media (Cresswell 1996). The response reflected the unease at the sight of militant women (active since the 1960s in groups such as the Women Strike for Peace movement in the USA) contesting the received realist wisdom of stationing nuclear weapons in a given location on the grounds that it would enhance the security of the population. In the UK and the USA, the early 1980s were dominated by the aggressive Cold War militarism of Prime Minister Thatcher and President Reagan, which led to the deployment of yet more nuclear missiles in Europe and the proposed development of an outer-space Strategic Defense Initiative (SDI, popularly known as Star Wars) designed to protect solely the United States from incoming missile attack from the Soviet Union.

For peace campaigners, America's development of the SDI in the early 1980s highlighted the global dangers posed by a nuclear confrontation and the extent to which nuclear arsenals were actually producing further insecurity and environmental degradation through hazardous waste production. Simultaneously, concerns over nuclear-power programmes had escalated following the Three Mile Island nuclear power station emergency in 1979 (USA) and the accidental destruction of the Chernobyl nuclear plant in 1986 (Ukraine). The meltdown of Chernobyl reactor number 4 resulted in the release of a massive dose of radiation into the local environment, which subsequently precipitated a transboundary movement of radioactive material into Western Europe. The 1950s optimism about cheap and safe nuclear energy was replaced in the 1980s by widespread fear over the safety of power plants and the storage of nuclear waste. A global risk society, as the German sociologist Ulrich Beck once noted, was materializing as the techno-scientific infrastructure associated with the nuclear bomb created uncertainties and insecurities.

In the South Pacific, the New Zealand government took the unusual step of acceding to the demands of anti-nuclear campaigners by declaring that American nuclear warships were no longer welcome in New Zealand territorial waters. It was a deeply controversial move because New Zealand and the United States had been close Cold War allies under various security arrangements. Nevertheless, in the 1980s the Wellington government began to question the wisdom of collective defence and refused to countenance nuclear weapons on their soil or in 'their' waters. Prime Minister David Lange became a widely cited and admired political leader within the anti-nuclear movement for his support of a South Pacific Zone of Peace and Co-operation.

American anger was considerable because as John Dorrance, a US consul general in Sydney, noted:

> There are those who fear the dangers of engagement. They are led to an understandable concern about the horrors of war to argue that a policy of non-involvement or isolationism is the way to save our countries from a nuclear holocaust. Unfortunately, we are all now physically within the reach of nuclear weapons and their secondary effects . . . the first requirement must be to maintain our collective strength to deter aggression (Dorrance 1985 cited in Dalby 1993: 217).

In contrast, Lange and peace campaigners argued that New Zealand's security was better served by developing peaceful relations with its neighbours in the South Pacific than by involvement in Cold War struggles dominated by the Northern Hemisphere. The cover of the 1987 Defence Review had a map of New Zealand constructed in such a way as to emphasize the proximity of the country to the South Pacific and Antarctica rather than the nuclear-armed Northern-Hemisphere states of the Soviet Union and the USA.

In the late 1980s, the New Zealand government and NGOs such as Greenpeace began to focus the world's attention on French nuclear testing. In conjunction with the island states of the South Pacific, Greenpeace orchestrated high-profile media campaigns with the aid of their ship *Rainbow Warrior* to interrupt proceedings at the French nuclear test site in Mururoa Atoll. The sinking of the *Rainbow Warrior* by two French secret agents in Auckland harbour in July 1985 (with the loss of one life) led to widespread condemnation and further rallied regional support. The 1987 Declaration of a Nuclear Free Zone in the South Pacific did not deter the French, and in September 1995 Greepeace relaunched its campaign against France when the latter resumed testing in Mururoa in the midst of negotiations in Geneva for a comprehensive nuclear test ban. Local political leaders condemned the French action as geographically and socially insensitive. Sir Geoffrey Henry, the prime minister of the Cook Islands, noted: 'I am prepared to accept that by some political concoction, the French have the right to test there, but geographically it is not theirs. It is part of the Pacific. It is as if an invasion has taken place' (cited in Dodds 2000: 108). Popular protest in the South Pacific culminated in rioting in Papeete, Tahiti (boosting support for the movement for independence in the French colony of New Caledonia) and demonstrations in New Zealand and Australia. In 1997 the French government finally announced that it would stop all forms of nuclear testing in the region as part of the NNPT ratification process. For those in the South Pacific, of course, the real cost of nuclear-weapons testing remains to be fully assessed.

In June 1996 at the South Pacific Forum, former New Zealand prime minister Jim Bolger proposed that the entire Southern Hemisphere be declared a NWFZ. In doing so, Bolger was echoing a widely held opinion in Australia, New Zealand and the South Pacific that the nuclear-weapons process was largely controlled and perpetuated by Northern-hemispheric nations testing their weapons in the allegedly 'empty South'. In order to advance the goal of a Southern NWFZ it was decided to work with the UN via a range of regional agencies and organizations such as the South Atlantic Zone of Peace and Co-operation, African Nuclear Weapon Free Zone, the Indian Ocean

Rim Initiative and the South Pacific Zone of Peace and Co-operation. This proposal appeared particularly poignant given that at this meeting of the South Pacific Forum, the Marshall Islands government reported on a proposal by the US government to create repositories for nuclear wastes on several atolls which had been contaminated by the 67 nuclear explosions carried out by the US between 1946 and 1958.

Nuclear nationalism in the 1990s: India and Pakistan

The non-proliferation of nuclear weapons has been a major priority for post-Cold War US foreign policy. In the midst of the Cold War, in 1968 the superpowers played a key role in formulating a Nuclear Non-Proliferation Treaty (NNPT) in an attempt to restrict the proliferation of NWS. With the collapse of the Soviet Union in 1991, the US funded a Co-operative Threat Reduction Programme (CTRP) to aid disarmament and denuclearization across the former USSR. Former Soviet republics such as Belarus, Kazakhstan and Ukraine were persuaded to give up their nuclear arsenals in 1991 (in return for financial credits and investment) as the US feared that other groups and states might seek to obtain these nuclear materials. The CTRP is also intended to assist the former Soviet Union in improving its storage facilities and monitoring the disintegrating nuclear capability of the former Soviet armed forces.

For many supporters of non-proliferation the NNPT remains the best hope for global disarmament because the leading nuclear powers agreed to abide by common rules: no NWS must transfer nuclear weapons to other states; non-NWS must not develop these weapons; and the IAEA must check that the NNPT is being properly observed. Key parts of the NNPT are Article IV, which allows NWS to develop technology for the peaceful use of nuclear energy, and Article VI, which commits NWS to pursue effective measures designed to promote global disarmament. Sceptics have argued that the aims of the NNPT, whilst laudable, are inadequate in terms of dealing with nuclear non-proliferation. The 1995 Nobel Peace Prize winner, Joseph Rotblat, commented that:

> The somewhat ambiguous wording of the Article [VI] has been cunningly exploited by the nuclear-weapons states to allow them to wiggle out of the obligation it imposes on them: one interpretation is that Article IV commits them only to pursue negotiations; another is that nuclear disarmament should be attempted only as part of a general and complete disarmament. However, the main purpose of the NPT is nuclear disarmament: the Preamble makes this quite clear. All signatories of the NPT accept certain obligations: the non-nuclear weapons states not to acquire nuclear weapons, and the nuclear-weapon states to get rid of theirs (cited in Huque 1997: 4).

One hundred and seventy-eight countries signed the NNPT by May 1995, but Israel, India and Pakistan refused. The dangers posed by nuclear proliferation are only too apparent in the South Asian region. China, as a major nuclear power, only formally acceded to the treaty in 1992. Since Independence in 1947, India and Pakistan have been locked in a bitter political and economic competition, which has extended to territorial conflict and nuclear-weapons proliferation. Ten years after the first Chinese

Figure 5.3   The Indo-Pakistani border near Amritsar. Rival security forces monitor movements closely and passage is strictly controlled.
Photo: Klaus Dodds

nuclear detonation, India was the first South Asian state to emulate this feat by secretly exploding a nuclear device in 1974, at the same time as Pakistan proposed to the United Nations that South Asia should be a NWFZ. Indian officials have consistently argued that the NNPT attempted to protect a privileged cartel of states, discriminating against non-nuclear Third World states. After the Indian nuclear event, the Pakistan government of the time committed itself to producing a nuclear bomb of its own, and in 1992 government officials confirmed that they now possessed a nuclear capability. During this period, both countries were enhancing their missile capability as Pakistan developed the Half-2 and India the Prithvi missile (Huque 1997: 3). Political rivalry, national prestige and projections of national power were widely held responsible for the massive investment in time and resources (Fig. 5.3).

India and Pakistan consider the NNPT to be discriminatory as it seeks to prevent Third World states from developing technologies enjoyed by the North for the last fifty years. The IAEA has also been accused of pursuing so-called threshold states such as Iraq with inspection visits while ignoring the activities of Israel and other Western allies. As the Indian representative to the UN, K. P. Unnikrishnan, noted in 1993: 'India could not subscribe to a Treaty or an attitude, which divides the world into haves and have-nots, with an inherently inequitable set of responsibilities and obligations of the two' (cited in Huque 1997: 11). In the absence of total disarmament, India declared a right to develop nuclear technologies regardless of the NNPT. American attempts to restrict the nuclear development programme in India have enjoyed only moderate success, in the main limited to imploring the Indian authorities to adhere to IAEA safeguards for nuclear installations. Pakistan has appeared to be more supportive of the NNPT, and since the 1970s has endorsed the idea of a nuclear-free zone for South Asia. However, both India and Pakistan have refused to sign the NNPT unless mutual agreement over nuclear weapons in the region and a joint signing of the treaty can be achieved.

The South Asian experience has exposed the real problems faced when trying to persuade sceptical Third World countries to reject nuclear-weapons technologies. American approaches to South Asian non-proliferation have often been contradictory: on the one hand, American technological and financial aid enabled both India and Pakistan to develop their weapons capability and on the other hand, according to Pakistan, American military assistance has favoured India in the sense that the US has carried out joint naval exercises with the latter. Moreover, the Americans recently prevented a shipment to Pakistan of 28 F-16 jet fighters long paid for by the Bhutto government. The prospects for regional disarmament in South Asia would be assisted by a more consistent policy from the USA, linking the regional and global disarmament processes (see Slater, Schultz and Dorr 1993).

The nuclear stakes were raised again in May 1998 when India's government led by Atal Behari Vajpayee announced that five nuclear tests had been carried out in the Thar Desert close to the Pakistani border. According to the prime minister, the tests were necessary because India's national security had to be assured in the light of the threat posed by Pakistan and China and the nuclear states of the Middle East. A formidable community of scientists, defence experts and political commentators gathered in New Dehli to celebrate the testing and to neutralize those who condemned the nuclear demonstrations. In the midst of the celebrations, few media and political commentators showed much concern for the prolonged dangers faced by the tribal population in the uranium mining belt of Jaduguda in South East Bihar. National security dictates and the precarious condition of Indo-Pakistani relations is cited as the reason for the continued necessity of testing (see W. Walker 1998). During the tense period May–June 1998, Hindu nationalists seized upon the nuclear testing programme as evidence of India's greatness and proposed that a temple be constructed to commemorate the site of the five nuclear explosions: a *shaktipeeth*.

As a direct result of these tests, the Pakistani government ordered similar nuclear testing to be carried out in June 1998. Despite American pleas for restraint, many South Asian observers saw the Pakistani tests as a necessary 'show of strength' (see Table 5.1).

Table 5.1   Conventional and nuclear arsenals: India and Pakistan

India	Pakistan
Defence budget (2003) $9.8 billion	Defence budget (2003) $1.5 billion
Armed forces 1.3 million	Armed forces 620,000
Nuclear forces	Nuclear forces
100–150 Agni II missiles	25–50 Haft V missiles
Range 2,000 km	Range 1,300 km

Data: International Institute for Strategic Studies (IISS), London

The American and other Western governments such as Germany retaliated by imposing sanctions on both India and Pakistan. The underlying problem of Indo-Pakistani relations is the inability of either side to transcend the narrow confines of a realist-defined 'military security' and conceive of a form of human security which crosses borders and particular national spaces.

Nuclear-weapons testing is just another element in a dangerous and escalating struggle over disputed territories in the Kashmir and the Punjab, which have witnessed frequent massacres and mutual bombing. Arguably, the so-called US 'war on terror' has provided a further strategic rationale for both governments to maintain high levels of defence spending, including on nuclear-weapons operations. India, for example, spends $1 billion per annum on nuclear-related matters. In May 2002, in the aftermath of acts of terror around the Line of Control, over 1 million troops gathered either side of the border. Despite the eventual avoidance of a widespread confrontation, newspapers in South Asia and elsewhere warned at the time that World War III was dangerously close to unfolding.

US, WMD and 'nuclear outlaws' in the post-September 11th era

American attempts at nuclear reduction have only been moderately successful and Washington's support for US–Russian joint reductions in 1992 and 1993 have to be counterbalanced with its reluctance to end all forms of nuclear testing. In 1995 the Americans proposed that even after the ratification of the Comprehensive Test Ban Treaty (negotiations began in 1994) there should be an opt-out clause after the first review conference. This was widely condemned and in 1997 all countries including other high-profile nuclear testers such as France and China endorsed the CTBT. For many smaller NWS such as India and North Korea, the reluctance of the Americans to commit themselves to total nuclear disarmament leaves the way open for others to follow suit until the original superpowers make the final commitment (see, more generally, Ayoob 1993, 1995).

This unwillingness to embrace a global regime of disarmament has arguably worsened in the last decade under the Clinton and Bush administrations. Reducing the ownership levels of WMD was never going to be straightforward for a country (the USA) that derived a great deal of national prestige and domestic economic importance from the Bomb. Successive US administrations have used defence spending to stimulate the domestic economy and many communities throughout the US remain dependent on the military and nuclear bases. During the Cold War, for example, the Strategic Air Command maintained 15 bomber bases, 9 ICBM stations and 3 submarine bases including the Whiteman Base in Missouri, which is home to the nuclear-capable B2 bombers. The latter were used during the 1991 Persian Gulf War, the 1999 bombing of Kosovo and Serbia, Afghanistan in 2001 and the 2003 assault on Iraq. Billions of dollars have been invested in a military and nuclear-weapons infrastructure, and politicians of all political persuasions are only too aware of the consequences of reducing those investments.

Internationally, the Clinton administration (1993–2001) confronted a world in which there were eight NWS and other states such as Iraq were known to possess a chemical and biological weapons capability including Sarin gas and VX nerve agents. The crumbling edifice of the Soviet Union revealed that the nuclear-weapons stock held by Soviet forces was in a deteriorating condition and concerns were expressed that nuclear materials could simply disappear in the chaotic aftermath of the 1991 break-up of the Soviet Union. Two major new sources of danger were identified: first, 'nuclear outlaw' states such as Iraq and North Korea which refused to allow IAEA inspectors to examine their WMD facilities, and second, WMD terrorism following the theft of nuclear materials from the former Soviet Union and evidence of isolated attacks by groups such as the Aum Shinrikyo cult, which in 1995 released Sarin gas in the Tokyo subway to deadly effect.

This new geography of WMD dangers has arguably heightened in the aftermath of the September 11th attacks on the US and the declaration by President George W. Bush (2001–) of an 'axis of evil' involving Iran, Iraq and North Korea (see Chapter 9). At the heart of these fears lies an uneasy tension. On the one hand, the dangers posed by WMD have always been global to the extent that the NWS had the capacity to inflict massive damage on the entire planet. On the other hand, the US and other states continue to identify certain states such as North Korea and Iraq as clear and present dangers while at the same time pursuing strategies such as the NMD (regardless of whether it actually works) which will cause fear and resentment among NWS including Russia, China and India (see In focus 5.4).

NMD may be attempting to create the ultimate 'gated community' but it also threatens to destroy the provisions of the 1972 Anti-Ballistic Missile Treaty. Confronting global dangers such as nuclear weapons requires mutual security and confidence-building mechanisms that are ultimately inclusive rather than exclusive. Technological solutions such as the NMD will not offer geopolitical protection from a state such as North Korea, which feels itself imperilled by a US nuclear-weapons arsenal and 38,000 conventional forces stationed in South Korea and/or close to the border zone. In October 2002 North Korea admitted that a covert uranium-enrichment programme existed and that

---

In focus 5.4: A new geography of threat in the aftermath of September 11th?

In March 2002 a Pentagon report listed seven nations which could be on the receiving end of a US pre-emptive nuclear strike. The unlucky seven were a mixture of NWS and non-NWS: China, Iran, Iraq, Libya, North Korea, Russia and Syria. The report also cited possible scenarios which might lead to such a confrontation going nuclear – North Korea invading South Korea, an assault on Taiwan by China and/or an Iraqi attack on Israel and/or Kuwait. Once published, the report was unsurprisingly criticized by all seven listed states as deeply provocative.

At the same time as the report was released, the US proposed to resume nuclear testing and continue the development of so-called 'mini-nukes'.

the 'axis of evil' speech by President Bush in February 2002 undermined earlier attempts to engage with a regime which had liaised with Iran and Pakistan over ballistic-missile technology (Cotton 2003). The Bush pre-emptive strategy is in danger of failing to confront the contribution made by the United States to the production and circulation of global (in)security.

Conclusions

This chapter has explicitly rejected the claim by some conservative commentators that nuclear proliferation may actually contribute to regional and global stability because it introduces a sense of caution into the decision-making processes of political and military leaders. Nuclear weapons are also fundamentally moral issues, often conflicting with realist views of international political life which stress national security and political self-interest (see White, Little and Smith 1997). WMD including nuclear weapons remain a striking symbol of the globalization of danger. In the aftermath of the Cold War, Western fears over nuclear, biological and chemical weapons have focused on India, Pakistan and Iraq – strategic environments in which realist-inspired geopolitical imaginations flourish.

The problems posed by nuclear weapons need to be located within a matrix which acknowledges that militarization is intimately connected to other concerns such as North–South relations, development, poverty, environmental problems and the construction of danger. Promoting processes of demilitarization requires us not only to consider how nuclear weapons can be reduced in number but also why states seek to retain WMD. Any future management of international nuclear behaviour will necessitate a careful analysis of global security which is sensitive to the ways in which threats and dangers are represented in the post-Cold War era. This has direct implications for the kinds of common security mechanisms that may emerge to confront a global problem rather than one simply located in 'axis of evil' states such as Iran, Iraq and North Korea. One terrible legacy of the 2003 assault on Iraq is that it may well have convinced other states that the possession of WMD is one of the best defences against intervention by the United States and its allies. Although Iraq was ostensibly invaded on the grounds of illegal possession of WMD, it is unlikely that the US would have been so confident about intervening militarily if it had really believed that Iraq had a fully operational nuclear-weapons capability.

Key questions

- Why did nuclear weapons proliferate in the post-1945 period?
- What attempts were made to control and/or resist nuclear-weapons proliferation?
- Why is the United States investing in NMD?
- Does the possession of WMD offer states the greatest protection against a pre-emptive strike from the United States or any other NWS?

- Does the United Nations need further institutional strengthening in order to confront the dangers posed by WMD proliferation?
- Will the implementation of the NMD by the United States improve or worsen global security and the control of WMD?

Further reading

R. J. Lifton and R. Falk, *Indefensible Weapons: The Political and Psychological Cases Against Nuclearism* (New York, Basic Books, 1982). J. Newhouse, *The Nuclear Age* (London: Michael Joseph, 1989) and M. Foot, *Dr Strangelove I Presume* (London, Michael Joseph, 1999). A short yet highly readable guide to nuclear culture is to be found in C. Newkey-Burden, *Nuclear Paranoia* (London, Pocket Essentials, 2003). For feminist arguments concerning nuclear weapons see J. Seager. *Earth Follies* (London, Earthscan, 1993) and J. Pettman, *Worlding Women* (London, Routledge, 1996). For critical geopolitical evaluations of nuclear tensions in the 1980s see S. Dalby, *Creating the Second Cold War* (London, Belhaven, 1990) and for the challenges confronting policy makers in the contemporary era see J. Cirincione, *Repairing the Regime: Preventing the Spread of WMD* (London, Routledge, 2000). On the relationship between the United States and the so-called 'nuclear outlaw' state of North Korea see J. Cotton, 'The second North Korean nuclear crisis', *Australian Journal of International Affairs* 57 (2003): 261–80. This relationship was of course addressed in part in the James Bond film *Die Another Day* (2002).

Websites

Campaign for Nuclear Disarmament www.cnduk.org
Department of Defense (USA) www.defense.mil
Pugwash Conferences on Science and World Affairs www.pugwash.org
Stockholm Peace Research Institute www.sipri.se

Chapter 6

GLOBALIZATION OF ENVIRONMENTAL ISSUES

Key issues

- Why do environmental issues such as global warming challenge the national sovereignty of states?
- How have environmental issues exposed inequalities between North and South?
- What did the international community and non-state organizations at the 1992 Rio Summit achieve?
- Rather than helping to solve problems, does the current interstate system alongside global capitalism merely produce more transboundary environmental issues?

In his book, *Resource Wars*, the American writer Michael Klare (2001) contends that disputes over resources (often located in fragile environments) will become increasingly prominent in global politics. He even identifies a basic geography of 'flashpoints' – regions such as the Caspian Sea, the South China Sea and the Middle East, which due to their resource potential (including scarce water supplies) could provoke a series of confrontations involving Great Powers, client and regional states, and multinational corporations (see In focus 6.1).

The United States government has been extremely active in the Caspian Sea region to the extent that in 1997 military personnel were dispatched to the Central Asian state of Kazakhstan in order to carry out simulated exercises involving Kazak and other regional forces. This operation was not launched for purely military reasons – the Caspian Basin has been identified by US military planners as one of the most significant alternative sources of energy outside the Middle East. It illustrates a form of practical geopolitics, as regions such as the Caspian Sea are defined on the basis of their resource potential and strategic access.

In the aftermath of the September 11th attacks on New York and Washington DC, the US Pentagon increased its military presence in Central Asia (for reasons such as its proximity to Afghanistan). In other words, access to Caspian Sea oil is being considered as a security issue rather than a challenge to existing modes of industrial behaviour and/or resource consumption.

Access to precious energy and/or mineral sources is only one part, however, of an increasingly global environmental equation involving rising population growth, loss of biodiversity, climate change, resource exploitation, widespread poverty, airborne

In focus 6.1: Resource flashpoints

1. **Caspian Sea Basin**. It is estimated that the untapped oil and natural gas resources of the region are sufficient to provide major energy consumers such as the US with an alternative source of supply to the Middle East. However, the ownership of resources is disputed between four states: Azerbaijan, Kazakhstan, Iran and Russia. There is also uncertainty as to how American corporations would export the oil from the region given that most of the pipelines currently in existence traverse Russia, which is deeply suspicious or even resentful of the US military presence in the former Soviet Union. Other states such as Turkey and China are also keen to extend their geopolitical and resource interests in the Caspian Sea basin.
2. **South China Sea**. It is widely suspected that oil and natural gas deposits are located close to the highly disputed Spratly Islands. These uninhabited islands are claimed by China, Malaysia, the Philippines, Taiwan and Vietnam. China is particularly active in its claim to the Spratly Islands, not least because its dependency on oil and natural gas is growing rapidly. Japan, which is hugely dependent on Persian Gulf oil, is watching with concern (because of its perceived resource needs) given that the South China Sea is a major transit area for those energy supplies.
3. **Middle East**. By 2040/2050, it is estimated that known oil resources in the region will be exhausted. The largest reserves are held by Saudi Arabia and Iraq. It is frequently argued that the removal of Saddam Hussein's regime in 2003 in combination with arms sales to other US-friendly Gulf states was motivated by a concern to preserve US access to those reserves.

The US Pentagon has conceptualized these three regions as part of a gigantic 'strategic triangle', which will shape the pattern of potential wars in the twenty-first century.

Further reading: *see* Klare 2001.

---

pollution, ozone depletion, coastal and oceanic pollution, forestry and indigenous plant loss, fossil-fuel consumption, and new information concerning the human impact on the environment. Resources, despite the claims of some realist writers, can never be divorced from other issues such as pollution and global equity. There is now a widespread sense in which the limits of social progress and industrial development are being reached; hence many realist writers fear that future international environmental relations will be dominated by disruption and violence as states seek to either preserve their environments (including their biodiversity) or access precious resources such as water and oil (see Princern and Finger 1994, Doyle 1998, Klare 2001).

Other scholars draw attention to environmental and developmental issues which have not only assumed a far higher political profile but also challenge the traditional realist assumptions associated with national sovereignty and international political co-operation (see Imber and Vogler 1995, Castree 2003, Dalby 2002, 2003 and Liftin 2003). How, they might ask, is military force going to help resolve global climate change? Why should we worry more about access to oil than about pollution? Is the interstate system the best mechanism for protecting or responding to environmental challenges such as global warming or biodiversity? As before, the kinds of questions we ask in the first place help to define the answers we might ultimately produce. There is no one correct way of looking at this issue or others such as humanitarian intervention (Chapter 7) or terrorism (Chapter 9). Growing worry over environmental degradation (local loss of biodiversity, for example) and the possible impact of global warming has been linked to contemporary debates over planetary security, global governance – even modernity itself (see Dalby 2003, Liftin 2003). Concern for the wider impact of environmental affairs was firmly consolidated at the United Nations Conference on Environment and Development held in Rio in June 1992 (known as the Rio Summit). More than a hundred governments and heads of state joined hundreds of NGOs to discuss the numerous challenges and issues facing the global ecosystem. Worldwide industrial and urban development has given rise to what we might be called 'planetary geopolitics', rather than merely 'international politics' (Liftin 2003).

Most critical geopolitical writers interested in environmental affairs recognize that ozone depletion, climate change and environmental degradation pose troubling questions for global politics because these issues are frequently transboundary and therefore beyond the controlling remit of any one state (see Shafer and Murphy 1998, Dalby 2002). Environmental implications have arguably helped to shatter confidence in the modern geopolitical view (see Chapter 2) based on the inviolability of national boundaries, the distinction between domestic and foreign, the growing influence of environmental NGOs and transnational corporations (TNCs) and finally international governmental organizations (IGOs) responsible for co-ordinating environmental matters (Castree 2003: 426 and more generally Agnew 2002). Boundaries have become highly vulnerable, sovereignty has been weakened and even ceded by states while increasing numbers of NGOs have become involved in the management of environmental matters and associated issues such as human migration and complex humanitarian disasters. But far from diminishing the claims of states and exclusive sovereignty, human survival arguably depends on the capacity of states and other organizations to collaborate in unprecedented ways in order to protect the earth's ecosystems.

Globalization theorists and writers from the Global South continue to stress that environmental issues have to be considered as part of a wider matrix concerning poverty, consumption, development and North–South relations (see Mittleman 1996, Spybey 1996, Thomas and Wilkins 1997, Held *et al.* 1999). In their 2003 Report on Human Development, the United Nations recognized that the world's poorest 20 per cent enjoyed only 2 per cent of the world's income (UN 2003). Whereas Northern countries worry about climate change and ozone depletion, Southern states are often more preoccupied with population increase, resource scarcity, basic needs and poverty reduction. Allow me to reiterate a fundamental dictum: our definitions of 'environmental issues' can often determine the sorts of analyses and policy options we produce (Lipschultz and Conca 1993, Hurrell 1995).

From Stockholm to Rio: transboundary and global agendas

Environmental concern has emerged as a significant feature of global politics since the late 1960s. This is not to claim, as Noel Castree reminds us, that environmental affairs were not important in an earlier era (Castree 2003: 424). Earlier patterns of recognition were not seen as fundamentally challenging the contemporary condition of world politics. The best-selling book by Rachel Carson entitled *Silent Spring* helped to mobilize public interest in environmental politics in Northern societies such as the United States and Western Europe (Miller 1995: 6). More specifically, the creation of non-state movements such as Greenpeace and Friends of the Earth in 1969 consolidated public interest in environmental management and conservation. Some environmental thinkers have argued that this concern was also indicative of a post-1945 generation anxious to warn the wider world of the dangers of modern industrial development and nuclear fallout (Dobson 1990). Media coverage of environmental disasters in the late 1960s such as the sinking of the *Torrey Canyon*, which involved the loss of 875,000 barrels of crude oil and polluted sections of the Cornish coastline, alerted the public to the dangers posed by massive oil spillages to the coastal environment (see In focus 6.2 and Fig. 6.1).

Since the mid-1960s, debates on international and/or transboundary pollution have occupied environmentalists and governments in the North and the South alike. There

Figure 6.1   Greenpeace and their protests against commercial logging in Siberia
Photo: PA Photos

In focus 6.2: Greenpeace

Greenpeace was founded in British Columbia, Canada in 1971, initially to oppose the US underwater nuclear testing in Alaska. The group's reputation for direct action was epitomized in that campaign, in which it sailed into the centre of the test site. Although unable to prevent the explosion, Greenpeace was successful in transforming a loose coalition of environmental writers, academics and sailors into a multinational organization by the mid-1980s. After the US government agreed to end underwater testing in 1972, the prevention of all further testing became Greenpeace's priority.

French nuclear testing was suspended in the South Pacific in 1975. When France renewed testing in the 1980s and 1990s, Greenpeace mobilized opposition and continued to campaign even after its vessel the *Rainbow Warrior* was sunk in Wellington harbour in 1985. Two French secret service agents were found guilty of this act of terrorism.

Achievements

The International Whaling Commission announced a moratorium on whaling in 1982 following Greenpeace's protests against the practice.

The 1983 London Dumping Convention established a moratorium on ocean dumping of radioactive waste following a Greenpeace campaign.

In 1987 Greenpeace established an Antarctic base in order to highlight its opposition to all proposals which would allow mining in the region. A new environmental protocol was established for Antarctica in 1991.

In 1995 Greenpeace prevented Shell from dumping the Brent Spar oil platform in the North Sea. The oil rig was later towed to Norway for dismantling on land. Disputes over the scientific data concerning the environmental impact of sea dumping lost Greenpeace a degree of credibility.

Through direct action and skilful use of media, Greenpeace has become one of the most effective environmental organizations, with membership in more than 25 countries and public subscriptions/donations of £30–40 million a year.

was concern not only that the effects of industrial pollution were placing dangerous pressures on state co-operation but also that the ecological limits of the earth had been reached. The transboundary nature of phenomena such as acid rain made it evident that local and regional environments could not be maintained or even protected by individual states or regional strategies. 'Acid rain' was a term adopted by media and environmental campaigners in the 1980s to highlight growing awareness of airborne chemical pollution (in the form of sulphur dioxide and other pollutants) and its effect on vegetation and the food chain. While scientists have known about the problems of industrial pollution for some time, it was only in the last decade that political leaders started to press for preventive action. Former British prime minister Margaret Thatcher acknowledged in April 1986 that UK industrial pollution was causing acid rain to fall on Scandinavian forests with devastating consequences (see In focus 6.3.).

Political scientists and geographers label this form of pollution as transboundary because acid rain can fall on areas many miles away from original sources such as factories, power stations and volcanoes. International agreement concerning the reduction of acid rain has been slow, with some states such as the USA and the UK unwilling to accept large-scale reductions called for by the Nordic countries.

A shift in attitude caused a change in focus from local problems of pollution and waste management to global issues of ecosystem management. In 1968 the Intergovernmental Conference of Experts on the Scientific Basis for Rational Use and Conservation of the Resources of the Biosphere met in Paris to discuss human impact on the biosphere, including issues such as overgrazing, deforestation and water pollution. The scale of human activity is such that many 'green' commentators and scientists are concerned that the maximum capacity of the biosphere to absorb and sustain such activity is rapidly reaching saturation point. High-profile reports such as *Limits to Growth* (1972) claimed that economic growth, in terms of increased production and consumption of goods and services, could not continue in an unchecked manner because of the implications for

---

In focus 6.3: The most polluted place on earth? Norilsk in Arctic Russia

Soviet leaders minimized concern for environmental issues, believing that communism could master the 'natural environment'. Large-scale dam construction, massive river-diversion schemes and mineral exploitation degraded environments across the Soviet Union. One of the most notable examples of this chronic degradation was the Arctic town of Norilsk, a remote mining town established during the era of the Soviet Union (1917–91). Despite the break-up of the Soviet Union in 1991 and a downturn in production, Norilsk's factories continue to emit 5,000 tonnes of sulphur dioxides into the atmosphere per year. Life expectancy is ten years lower than in the rest of Russia, with the Norilsk Mining Company producing one seventh of all the factory pollution in Russia. The company is accused of not only interfering with local environments but also affecting Norwegian and Canadian ecosystems through airborne pollution. Norilsk remains a 'closed town' which foreigners are banned from visiting because of sensitivities around the mining operations. Such operations would have to be severely curtailed if Russia were to ratify the Kyoto Protocol.

This illustrates how the local economic and environmental circumstances of one town in Arctic Russia can have implications for other peoples and places as far afield as Canada and Scandinavia. Current forms of resource exploitation carry considerable ecological costs, yet an international agreement (such as the Kyoto Protocol) may not be ratified by the Russian government because without costly investment to reduce emissions industrial output would be compromised.

Source: Paton Walsh, 2003

future environmental management. The metaphors employed by the neo-Malthusian writers of the 1970s portrayed the earth as a 'lifeboat' and/or as a spaceship to highlight its ecological capacity to handle pollution and economic growth. While this corpus of literature was often criticized and condemned for failing to acknowledge North–South inequalities and the social mechanics of life, it did acknowledge the limits to development (see Doyle 1998, Doyle and McEachern 1998).

The 1972 UN-sponsored Conference on Human Development in Stockholm sought to explore further some of the themes addressed at the 1968 intergovernmental meeting in Paris. In contrast to the Paris meeting, the Stockholm conference embraced the political, economic and social issues connected to human development and attracted states and NGOs. The lobbying by the Third World political coalition, the G77, ensured that the Stockholm agenda considered issues such as water supply, poverty and shelter, in order to broaden Northern concerns over population growth, resource exploitation and limits to economic growth (Miller 1995: 8, Williams 1997). The G77 also called for an acknowledgement of the links between welfare, industrial development and environmental degradation. For the first time in a public forum, differences between North and South over environmental issues became abundantly obvious (see Chapter 3). Three major issues raised by G77 delegates were: Who was responsible for environmental degradation? Did the South have a right to develop along the same lines as the North? And should the North offer the free transfer of 'clean technology' to the South?

After the 1972 Stockholm Conference on the Human Environment, the UN adopted a more active role by furthering ozone-layer protection (Montreal Protocol, 1987), regulating the disposal of hazardous waste (Basel Convention, 1989), establishing rights and responsibilities for the oceans (Law of the Sea, 1982), highlighting the problem of overfishing by producing reports on the world's fisheries by the Food and Agricultural Organization (FAO), and limiting further tropical deforestation through the Tropical Forest Action Plan (Castree 2003: 429). These achievements raise two fundamental issues: first, could states actually respond effectively to transboundary challenges by constructing new agreements; and second, did states and their governments agree about the definition of 'environmental problems'?

The United Nations Environment Programme (UNEP) was created in the aftermath of the 1972 conference for the purpose of addressing global environmental issues and North–South relations. Over the next 20 years environmental and economic issues were debated, ranging from ozone depletion, forestry and sustainable development to climate change. Northern states such as the USA have concentrated their diplomatic energies on ozone depletion while Southern states such as India have sought to draw wider connections between environmental destruction, poverty, debt and development. Although the G77 had a limited impact on the outcome of these negotiations, it provided a forum for Southern accusations that Northern countries were reluctant to acknowledge responsibility for most of the damage to the earth's biosphere caused by industrial activities and/or that Northern states were now trying to curtail the developmental aspirations of Southern states. The Indian writer Vandana Shiva referred to this situation as a form of 'ecological imperialism' whereby the North seeks to instruct the South on how to reform its industrial behaviour whilst refusing to assist in debt reduction and/or technology transfer (see Shiva 1993).

The link between North–South relations and environmental issues was recognized in the Brundtland Report of 1987. By using the term 'sustainable development', it was explicitly acknowledged that the developmental needs of the South could not be marginalized by the Northern agendas of global environmental protection. However, throughout the 1980s the Northern bloc of the USA, Europe and Japan was unwilling to make concessions over industrial development and the consumption of resources (and hence sought to protect national sovereignties) while at the same time the South, led by countries such as India, China and Brazil, refused to concede its right to determine its development priorities in the face of evidence that the North (25 per cent of the world's population) consumed 70 per cent of the world's energy, 75 per cent of the world's metals and 60 per cent of total global food production. The problem facing many of the negotiators at major international conferences on climate change, ozone depletion and global warming was that no consensus over the nature of environmental issues, the meaning of sustainable development and the core principles of management for the future could be agreed. During the 1980s, therefore, environmental groups argued that states frequently committed themselves to non-binding conventions, which respected their sovereign interests at the expense of developing global and politically inclusive forms of protection for the environment.

Under the guise of sustainable development, environmentalists have argued that major states such as the USA and multinational corporations have sought to project a particular vision of sustainable development which privileges the capacity of the market and industrial development to produce ecologically friendly economic growth (see Redclift 1987). The representation of environmental problems is profoundly important in shaping subsequent debates and policy options. In this case, the 'limits to growth arguments' have been dispensed with because the increased efficiency of industrial farming and other production systems can effectively bypass these ecological limits. As Larry Summers, a former World Bank economist, noted in 1991: 'There are . . . no limits to the carrying capacity of the earth that are likely to bind any time in the foreseeable future. There is not a risk of any apocalypse due to global warming or anything else. The idea that we should put limits to growth, because of some natural limit, is a profound error' (cited in Seager 1993: 134).

For ecologists and globalization theorists, this unproblematic vision of the future is deeply troubling because it reduces environmental issues to questions of efficiency and effective planning, rather than recognizing that some profound moral and political issues are raised by industrial development and economic growth (see Porter and Brown 1996). As Michael Redclift noted in his oft-cited critique:

> Sustainable development, if it is to be an alternative to unsustainable development,
> should imply a break with the linear model of growth and accumulation that ultimately
> undermines the planet's life support systems. Development is too closely associated in our
> minds with what has occurred in Western capitalist societies in the past, and a handful of
> peripheral capitialist societies today (Redclift 1987: 4).

The 1992 Rio Summit was intended to be a forum for discussion and debate of these controversies, as well as a focus for moral pressure on governments across the globe while at the same time strengthening the ongoing work of local NGOs, women's groups and community-based organizations (CBOs).

Rio Summit and global ecology

One hundred and seventy states and their representatives, thousands of NGOs and many multinational corporations attended the 1992 Rio Summit on the twentieth anniversary of the 1972 UN Stockholm Conference. The purpose of the conference was to consider the environmental consequences of human development. Five years earlier, the 1987 Brundtland Report had warned that traditional patterns of economic growth were not sustainable in the long term, given the demands of the South for further industrial development. What became apparent at the Rio Summit was that Northern and Southern states were pursuing in the main different environmental agendas as the former were concerned with ozone depletion and global warming whereas the latter were anxious to address the relationship between economic development and environmental management. The Rio Summit produced conventions dealing with climate change, biodiversity, forestry and Agenda 21, but considerable differences and difficulties over the design and implementation of sustainable forms of development remained unresolved (see In focus 6.4).

In focus 6.4: National sovereignty and global environmental management

The transboundary nature of many environmental issues poses challenges for the contemporary interstate system. The question of how one balances national interests with a concern for the global ecosystem remains vexing. Realists would, by and large, contend that the nation-state must put its own national interests first and then collaborate where appropriate over issues such as global climate change. However, if the national interest is threatened then co-operation should be strictly limited.

Ecologists and critical geopolitical writers argue that there are two types of ecological challenge which have implications for conventional understandings of international politics. These are, first, transnational problems such as global warming and second, 'local problems' that have extra-local implications such as the loss of biodiversity. In the case of the latter, the loss of biodiversity would have serious implications for the exploitation of commercial crops and the development of pharmaceuticals.

These fundamental differences bedevil global environmental negotiations, not least because developing Third World states such as India and China argue passionately that the North is attempting to invoke 'global' priorities at the exact moment when other states wish to fulfil their national development potential.

The main achievement of the Rio Summit was to convene a global forum for the discussion of global environmental problems in the wake of UN Resolution 44/228, which called for such a meeting in 1989. This was no mean achievement given the profound differences of opinion that existed amongst the interstate community and environmental NGOs.

After weeks of preparatory meetings, the attending governments agreed to the following: 27 core principles of development and the environment; conventions on biodiversity and climate change; Agenda 21 and a host of other environmental agreements such as the creation of a Commission on Sustainable Development to help the UN monitor environmental progress. The main document, Agenda 21, declared in Article 1 that human beings were central to sustainable development. Article 2, however, reiterates that states enjoy the right to exploit their own resources. Underlying this document is a powerful commitment to upholding the right of states to decide their own environmental strategies even though it is acknowledged that states should seek to act in a sustainable manner. Indeed Article 15, the so-called precautionary principle, urged that: 'In order to protect the environment, the precautionary approach shall be widely applied by states according to their capabilities. Where there are threats of serious or irreversible damage, lack of scientific uncertainty shall not be used as a reason for postponing cost-effective measures to protect environmental degradation.' Although this sounds laudable, it underestimates the difficulty of balancing the future needs of humanity with the sovereign interests of states and the business interests of multinationals when scientific information is being challenged in an attempt to halt binding environmental agreements relating to industrial development.

In his critique of the Rio Summit, Tim Doyle accuses Northern political elites and TNCs of defining the ethos and content of Agenda 21 (Doyle 1998, Doyle and McEachern 1998). While environmental problems were defined in terms of global ecology, the problems of global warming and population growth were frequently discussed in Northern elite and scientific terms which marginalized the major environmental issues defined by the people and states of the South. Doyle argues that Agenda 21 perpetuated a form of sustainable development which continues to promote the goals of economic growth and industrial development through market liberalization and world economic regulation. The environment is viewed as a resource which can be used efficiently by particular human 'users' rather than as a fragile ecosystem whose fate is intimately bound up with that of all human beings. Instead of promoting profound change in human behaviour, the Rio Summit effectively approved existing forms of industrial development and outlined an approach for piecemeal change and legislation.

Third World countries which noted that Northern states were not willing to alter existing global systems of trade, finance and debt collection felt a keen sense of anger and disappointment with the Rio Summit. In 1994, a Global Conference on the Sustainable Development of Small Island Developing States was held in Barbados. Representatives from over 100 states attended the conference to consider the economic and environmental problems faced by small island states. The delegates approved a Programme of Action which called for measures to protect them from rising sea levels, the loss of natural resources and dependency on a few primary exports. The concerns of small island states such as the Maldives indicate that there is no 'Southern' consensus on the likely impact of climate change and that the criteria for judging such challenges vary considerably. The Maldives could, for example, disappear if sea levels were to rise in the next century due to ice-cap melting; hence agreements concerning global warming are of particular interest to them (see Chaturvedi 1998). Thus the conference revealed only too clearly that not all states and communities share the same priorities.

Environmental issues and sustainable development need to be considered in alliance with negative equity and net resource flows from South to North. The Indian ecological writer Vandana Shiva noted:

> The 'global' in the dominant discourse is the political space in which a particular dominant local seeks global control, and frees itself of local, national and international restraints. The global does not represent the universal human interest, it represents a particular local and parochial interest which has been globalised through the scope of its reach. The seven most powerful countries, the G-7, dictate global affairs, but the interests that guide them remain narrow, local and parochial (Shiva 1993: 149–50).

Disappointingly, the Rio Summit failed to address some of the most pressing problems facing global environmental politics, such as securing firm and binding commitments to cut carbon-dioxide emissions, reversing the militarization of the environment and imposing firm controls on the activities of multinationals (see Table 6.1).

As with most of the conventions negotiated at Rio, the Climate Change Convention was replete with ambiguities, omissions and qualifications to allegedly binding agreements. The problem of Third World debt and its link to poverty and maldevelopment was not considered, even though the alternative Global Forum had called for a greater willingness on the part of Northern states and banks to grant substantial debt relief and to promote the involvement of non-state organizations in the production of key documents such as Agenda 21.

The trend to privilege the role of the state and the interests of Northern multinational organizations has continued in the years following Rio. In July 1996 the United Nations Development Programme (UNDP) issued another report on sustainable

Table 6.1   Who are the largest polluters in the world?

Country	CO_2 emissions (millions of tons per year)
United States	5,410
European Union	3,171
China	2,893
Russia	1,416
Japan	1,128
India	908

Note: Figures based on 2001 estimates. The EU at that stage compromised 15 nation-states
Data: Steger 2003: 90

development, which actually called for further industrial growth in order to tackle the inequalities between North and South. Organizations such as the WTO and the Multilateral Agreement on Investment (MAI) are designed to remove any impediments to the global market economy. The MAI was negotiated between 1995 and 1998 and will legally bind states and limit their power to impose conditions and requirements on multinational investors, thus limiting the leeway to control trade and investment flows in particular national territories. Unfettered and open economic growth may well contribute to enhanced environmental degradation and could effectively weaken the capacity of states or non-state movements to counteract or even protect specific environments (see Herod, O Tuathail and Roberts 1998, Pilger 1998). Environmental issues are intrinsically linked to other concerns such as trade and finance.

In the future, it might be necessary to consider new ways to combat environmental challenges instead of delegating this responsibility solely to nation-states. Some believe that it will be necessary in the near future to create a World Environmental Organization (WEO) in order to counterbalance the work of the WTO. But that would merely add yet another international regulatory body composed of individual states. This is not intended to belittle the achievements secured by states, such as the Montreal Protocol in 1987 and the creation of organizations such as the Commission on Sustainable Development and the Global Environmental Facility. Many more states are, at least in principle, committed to sustainable forms of development now than they were 30 years ago. The interaction between national governments, TNCs and international organizations such as the World Bank and the WTO has implications for all our environments and political cultures.

North–South relations and the protection of the global commons

Areas of the world, which are not claimed by any one nation-state are referred to as *res communis humanitas*, and these include the earth's atmosphere, Antarctica, the ocean floors and outer space. The environmental protection of global commons is problematic due to the limits of interstate co-operation and the North–South divide (see In focus 6.5).

The protection of global commons places responsibility on the present generation to consider the needs and expectations of future humanity. The notion of a 'global common' and/or a 'common heritage of mankind' has been employed by the international community to signify regions which are not subject to the sovereign jurisdiction of the state. These are areas which by their very nature entail common managerial concerns. The question of responsibility for these areas remains undecided, given that the sovereign rights of states extend to the margins of the Antarctic, territorial waters and air space (see Vogler and Imber 1995, Vogler 1999).

Advances in our technological and scientific ability to exploit and degrade environments such as the ocean floors and outer space became increasingly politicized after the 1950s. The First UN Conference on the Law of the Sea established that coastal

In focus 6.5: Protecting Antarctica

Since 1959, the Antarctic Treaty (which has been signed and ratified by over 40 states including the United States, India, China, Russia and much of the European Union) has preserved the polar continent as a zone of peace, a continent for science and an environmental wilderness. All forms of military activity are banned inside the Antarctic Treaty zone and all parties commit themselves to preserving the only continent without an indigenous human population. The 1959 Treaty is essential because seven countries press territorial claims to the Antarctic (Argentina, Australia, Chile, France, New Zealand, Norway and United Kingdom) while others including the United States dispute the legal validity of those claims. By focusing on science, peace and environmental conservation, the treaty was able to 'sidetrack' this potential source of dispute.

This does not mean, however, that all interested parties have uncritically accepted the Antarctic Treaty System (ATS). Third World states such as Malaysia advocated in the 1980s that the United Nations should be in control of the Antarctic, especially if the continent's mineral resources were ever exploited. NGOs such as Greenpeace complained that the ATS failed to take into account the views and interests of non-state organizations, especially with regard to environmental protection. Tour operators wanted greater freedom to pursue commercial activities without restrictions being imposed by the ATS.

The entry into force of the Environmental Protocol in 1998 as an addition to the 1959 Antarctic Treaty helped to allay some of these concerns. Fundamentally, it banned all forms of mineral exploitation and placed the environmental protection of Antarctica at the heart of all future activities. Critics such as Malaysia have been far more sympathetic to the ATS now that there is no question of powerful states such as the US unilaterally exploiting the suspected mineral riches of Antarctica. While the question of sovereignty in Antarctica remains unresolved, the ATS has shown itself to be an international regime, capable of taking into account the interests of member states as well as non-member states, NGOs and commercial organizations such as tour operators. It is arguably the most successful international regime in existence.

states could declare exclusive rights over the adjacent continental shelf. Ownership of the ocean floor excited much international debate when the significance of these deliberations in relation to fishing and commerce became evident. New technologies such as oil and gas drilling, coupled with the effects of marine pollution, created an added sense of urgency. There was a widespread awareness that the oceans and seas were now even more vulnerable to the development of international economic enterprises.

The Third United Nations Conference on the Law of the Sea in 1982 extended this process by including new privileges and rights to the resources of the continental margin (Vogler 1999). By 1992, after a period of 30 years of negotiation, 154 states had agreed that the ocean floors and the sea could be incorporated into state ownership on the basis of a declaration of a territorial sea (up to 12 miles from the coastline) and/or an Exclusive Economic Zone (up to 200 miles from the coastline). Each coastal state could claim, under the 1982 UNCLOS III Convention, a 200-mile zone for the purpose of exploration, exploitation, conservation and the management of resources in the sea, seabed and subsoil (see Fig. 6.2).

This process of delimiting ownership of the waters and oceans has been highly unequal because more than half of the world's EEZs belong to 10 countries, including most of

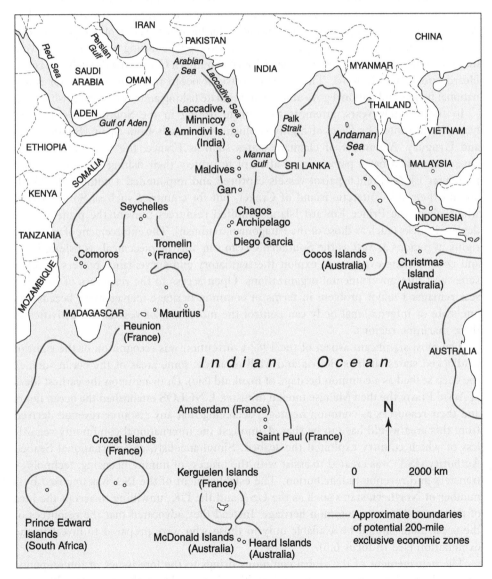

Figure 6.2   The Indian Ocean
Source: Adapted from Glassner 1996

the Northern states such as the USA, the UK, Japan, Russia and Canada. The biggest gainers in terms of submarine petroleum rights were Norway, the UK, the USA, Russia and Australia.

The Law of the Sea appears to favour a select number of Northern states, but enforcement of rights to the sea has become increasingly problematic. In the 1990s, the Patagonian toothfish, highly valued by American and Japanese consumers and fishing companies, became one of the most overharvested fish in the Southern Ocean (see Dodds 2000). Overfishing in this massive oceanic zone has been a problem in the past, resulting in the severe reduction in species numbers of the marbled rockcod and icefish. The Antarctic Treaty parties (through the Convention for the Conservation of Antarctic Marine Living Resources, in force since 1982) attempted to regulate fishing but failed to prevent illegal fishing in and around the various EEZs of Southern Ocean islands such as Crozet, Heard and Prince Edward. The sheer expanse of ocean makes it difficult for international bodies and national governments to regulate fishing and/or conserve fish stocks.

In the last few years, attempts to manage fishing in the Southern Ocean have been circumvented by fleets of highly sophisticated vessels from Spain, South Korea and Uruguay. A number of claimant states such as France, the UK and Australia have despatched naval patrols in an attempt to protect their fishing resources within particular EEZs. French patrol vessels captured and impounded a number of vessels around their sub-Antarctic island of Crozet, but for countries such as South Africa, which claim the Prince Edward Islands, limited resources prevent the protection of depleted stocks (such as those of the Patagonian toothfish). The enforcement of maritime rights in regions as vast as the Southern Ocean can only be piecemeal, as fishing fleets and governments expose and exploit the regulatory and surveillance powers of other states and intergovernmental organizations. Open access to the resources of the high seas remains a major problem in terms of common heritage management because no one state or international body can control the movement of vessels and activities in these maritime regions.

The most significant aspect of the 1982 Convention was recognition of the right of landlocked states such as Bolivia and Mali to access some areas of the ocean such as the deep seabed as a common heritage of mankind (sic). Drawing upon the earliest ideas of Arvid Pravo, the then Maltese foreign minister, UNCLOS established the ocean floors and their resources as common heritage, decreeing that any resource revenue derived from this area would have to be shared amongst the international community regardless of which country exploited the seabed. Simultaneously, an International Seabed Authority (ISA) was created to assist with the process of mining licensing, technology transfers and revenue redistribution. The establishment of the ISA was opposed by a number of Northern states such as the USA and the UK, unwilling to accept the idea of the ocean floor as a common heritage. Instead they advocated that the resources of the ocean floor should be available only to those who were prepared to invest in the exploitation (see In focus 6.6).

The management of the global commons brings to the fore issues of interdependence, vulnerability, and economic and political justice. Throughout the last 30 years, calls for global environmental protection have had to co-exist with demands for a NIEO

In focus 6.6: 'Problem solving' international regimes: the Antarctic Treaty System

It has been recognized, sometimes reluctantly as realists would contend, that environmental and resource management requires states to co-operate with one another precisely because pollution, people and resources such as fish transcend political boundaries. In order to promote co-operation, international regimes have been developed. A regime has been defined as a 'set of implicit or explicit principle, norms, rules and decision making procedures around which actors' expectations converge in a given area of international affairs'. In the case of the Antarctic Treaty System (ATS), for example, the following elements could be identified:

Principles: Science is seen to be the bedrock of the ATS because it has fostered a spirit of peace, co-operation and exchange. Environmental protection is deemed essential to the future management of Antarctica.
Norms: Freedom to carry out scientific investigation regardless of location and/or prior territorial claims is considered to be one of the most important features of the ATS and its management of the polar continent.
Rules: The 1959 Antarctic Treaty and the 1991 Protocol on Environmental Protection provide stipulations for the conduct of scientific and other forms of activities in the region.
Decision-making procedures: The annual meeting of the Antarctic Treaty Consultative Parties is the main forum for decision-making within the ATS. This is supplemented by a host of other meetings and conventions throughout the year.

To maintain the status quo, the member states have had to accept the principle of consensus. In other words, no major decision is taken unless all the parties accept the outcome. This means, much to the frustration of NGOs and activists, that decision-making can be slow as states seek to preserve the consensus.

Source: Krasner 1993, pp. 1–19

and territorial and resource sovereignties even extending to areas such as outer space, where claims to sovereignty would appear to be ridiculous. This area is riddled with North–South inequalities. As with the ocean floors and the earth's atmosphere, the crucial issue is technology and industrial development. It has to be recognized that the benefits of exploitative activities are highly unequal either for the present or future generations and it should also be remembered that it was representatives of the South, rather than the Northern states, who pressed for concepts such as common heritage to be applied to the global commons (see Chapter 3). Over the last 30 years, clashes over the exploitation of resources, the use of satellites and the emission levels of gases have been severe enough to suggest that co-operation has been limited rather than substantially enhanced since the 1950s.

Co-operation over the global commons will remain deeply problematic in spite of ground-breaking agreements such as the 1959 Antarctic Treaty, the 1987 Montreal

Protocol and the 1991 Protocol on Environmental Protection (see Paterson 1996). One of the enduring problems affecting all ecological issues is the difficulty of devising common environmental agendas on the basis of vague scientific evidence and time frames. In that sense the discovery of the ozone hole over the Antarctic in the mid-1980s revealed the opportunities and dangers inherent in tackling global environmental change (see Stokke and Vidas 1996, Dodds 1997). The British Antarctic Survey closely monitored the destruction of the stratospheric ozone layer in the 1980s because of fears that the release of chlorofluorocarbons (CFCs) was responsible for ozone reduction in the polar vortex over the Antarctic continent. As a region far removed from economic activity and population centres, the Antarctic was considered to offer an early warning of impending global damage to the environment. Prompted by this evidence from the Antarctic, the World Meteorological Organization and the United Nations Environment Programme (UNEP) set up an Inter-governmental Panel on Climate Change (IPCC) in 1988 with the aim of creating a forum for discussions and exchanges of scientific information on global warming.

The ethical challenges posed by the global commons were intensified in the aftermath of the 1992 Rio Summit, which, following the advice of the IPCC, created a Framework Convention on Climate Change (see Paterson 1996, Bush and Harvey 1997). While there was a strongly shared view that the world faced a series of environmental crises, there was only a weak agreement on the formation of and responsibility for joint action programmes. The Framework Convention was undoubtedly weakened by the unwillingness of the USA to support a technological and financial transfer to the South in order to promote clean technologies for industrial development. The then US administration also campaigned for a policy on greenhouse-gas emissions which allowed for reduction trade-offs, thereby undermining earlier commitments to the Montreal Protocol in 1987. Some states have clearly been more proactive than others in tackling climatic change. Low-lying countries such as the Comoros Islands and the Maldives (part of the Alliance of Small Island States) have a pressing interest in globally binding agreements for greenhouse-gas emission reductions.

Since the 1992 Rio Summit, progress in terms of developing binding environmental conventions dealing with climatic change and biodiversity has remained slow. The position is not likely to be improved rapidly, as environmental changes often appear gradually and because governments and other interested parties are apt to question available scientific evidence (Vogler 1995, Graham 1996). The Global Climate Coalition, for example, funded by the oil, natural gas and car lobbies, challenges existing evidence relating to global warming patterns and warns that energy bills would rise considerably if measures were taken to reduce carbon-dioxide emissions. Major events such as the Framework Convention on Climatic Change eventually agreed and formulated a resolution on the basis of what might be politically possible rather than what might be needed in order to cut pollution levels and halt the exploitation of the environment. It is precisely this kind of compromise that so frustrates NGOs who feel that important environmental issues are being sidelined for the sake of preserving international political agreement.

In spite of efforts to reform the consumption patterns of the North, such as the banning of CFC-containing products, the real battleground in terms of global environmental change is likely to be between North and South. As Andrew Hurrell has noted, global environmental change 'is an inherently global issue both because of the high levels of economic interdependence that exists within many parts of the world economy and because it raises fundamental questions concerning the distribution of wealth, power and resources between North and South' (Hurrell 1995: 131). For the South, major economic nations such as China, India and Indonesia argue that their developmental priorities have to be balanced with all calls for global environmental protection (Breslin 1996). Successive Chinese governments have been deeply suspicious of what they perceive to be the North's environmental agenda at a time when China's economy is growing in output and foreign direct investment. Large Southern economies have been unwilling to compromise their comparative advantage, although often with dire consequences for the environment. Moreover, as the recent logging-related forest fires in South East Asia and tropical hardwood exports from Africa have demonstrated, Northern multinationals may also be partially responsible for Southern environmental degradation.

This fundamental clash between North and South became evident once again during the negotiations leading up to the Kyoto Summit on global warming in December 1997 and in Buenos Aires in November 1998. The Clinton administration was accused of failing to lead the way, by its reluctance to commit the USA to a binding reduction in emission rates (currently 20 per cent of the world's total carbon-dioxide emissions) in a set period of time. This prompted the G77 bloc to reject calls made in September 1997 to establish binding targets for the South. Oil producers such as Saudi Arabia have also demanded that Southern states dependent on oil and natural-gas exports should receive compensation from the international community. Such a move, which was bitterly opposed by the USA, amounts to a demand for a NIEO in which the North should be prepared to compensate for the consequences of global environmental reforms in the South.

The fragmentation of Northern consensus on global environmental issues revealed fundamental differences over global warming between the US government and the European Union. The latter has argued that the North will have to demonstrate a firm commitment to emission reduction before persuading the South to follow suit. EU governments have accepted that the roots of many of the world's environmental problems lie with the North rather than the South. The European Commission in October 1997 announced that EU emission levels for the year 2000 would be 15 per cent below 1990 figures. This figure was eventually achieved and the UK's shift from coal to gas-fired power stations in the 1980s and 1990s has contributed to this process by allowing the UK to claim a 6 per cent decline in greenhouse emissions. The EU's plans to further reduce European emission levels will, however, be tied to agreements with Japan and the USA committing these countries to achieve similar reductions. As in the case of the 1992 Rio Summit, the US government remains reluctant to establish firm target figures in the light of lobbying from industrial and congressional sources which urged the administration of President Clinton not to threaten US jobs and profits margins. In Buenos Aires in November 1998 the US delegation agreed to cut emission rates more modestly.

The George W. Bush administration decided not to seek ratification of the Kyoto Protocol from the US Senate because it was widely suspected that there would be a limited constituency of political support. However, it also has to be acknowledged that the Bush presidency is heavily backed and implicated in the concerns of the fossil-fuel industry, which continues to dispute the claims of the IPCC that global temperatures are being affected by human behaviour. As the largest polluter in the world, the US is critical to the overall success of industrial emission reduction, and its unwillingness to ratify Kyoto has dismayed EU states and the Climate Action Network of NGOs. Advocates of the Kyoto Protocol also point to the manner in which the United States sought to take advantage of so-called 'carbon sinks' such as forests and 'pollution credits' from other countries in order to offset still further their own levels of industrial pollution (Paterson 2000).

What does this approach to the problem of global warming tell us about the George W. Bush administration's geopolitical worldview? For one thing, it reaffirms the realists' argument that national security (in the form of protecting domestic economic interests) remains the highest priority of this particular government. It also suggests that powerful states like the US and Russia are often adept at exploiting international agreements for their own benefit. In my view, any proposals for the protection of the earth's atmosphere and Antarctica have to acknowledge not only the unequal power relations between North and South but also the unilateral action of powerful states such as the United States and large polluters such as Russia. Global agreements such as the Climate Change Convention are often flawed in the sense that proposals to cut emissions can overwhelmingly reflect Western scientific assessments and environmental values (see Gupta 1997). Third World critics have frequently asked for some distinction to be made between the 'luxury emissions' of the North and the 'survival emissions' of the South. The role of the United States (and multinational industries involving cars, steel and power production) in rejecting the provisions of the Kyoto Protocol also reminds us that states have very different capacities and agendas.

The management of climate change is not, therefore, just a 'technical' issue. It also touches upon questions of political economy and international political influence as expressed either unilaterally or through international organizations such as the United Nations. If Sudan or Nepal unilaterally pulled out of the climate change negotiations, few would be unduly concerned. In other contexts, however, the environmental protection of the Antarctic is determined primarily by the US, the UK, the European Union, New Zealand and Australia, not only in terms of their capacity to influence the decision-making processes of particular international regimes but also through their sheer environmental impact in terms of industrial emissions. Northern NGOs such as Greenpeace have therefore been important advocates of more appropriate, cautious and precautionary action and behaviour in the wake of these ever-widening North–South inequalities. Moreover, they and their Southern counterparts have also drawn renewed attention to the prevailing politics of power and knowledge.

Conclusions

Global politics relating to environmental concerns call into question: the capacity of states to deal with climate warming; the ability of the international arena to facilitate co-operation; the willingness of environmental movements and TNCs to propagate region-ally sensitive policy options and strategies; and the role of international institutions and regimes in contributing to a wider culture of obligation. The inability of inter-national regimes to sanction action against states and organizations which fail to meet particular environmental standards is a worrying problem. Who will cajole China, Russia or the USA, for instance, if they fail to adhere to their carbon-dioxide emission quotas or simply opt out of the entire negotiating process? The answer may lie in a coalition of states and organizations including NGOs equipped to exert pressure and adept at 'shaming' parties who fail to meet their international obligations. Television and other forms of media networks (in combination with international and NGO agitation) might also further environmental action by exposing wrongdoing on the part of states and multinational corporations.

Even in the aftermath of September 11th, environmental issues remain a salient feature of global political agendas. President Bush's rejection of the Kyoto Protocol does not mean that global climate-change negotiations are fatally undermined. Kyoto was never intended to be the end point of negotiations. Given the potential scale and significance of climate change, negotiations will have to continue over generations. Opponents of the Kyoto Protocol will have to acknowledge that the risk and uncer-tainty associated with climate change, the loss of biodiversity and the management of the global commons will have implications for even the most powerful states in the world.

In terms of the future of the planet, it is not at all clear whether 'problem-solving' international regimes and their state sponsors alone are sufficient to protect future generations (including flora and fauna). Arguably, all conventions and treaties achieve is a response to contemporary policy problems. More fundamental questions such as why we face such substantial environmental challenges would require us to make a critical (and politically unpopular) appraisal of the kinds of societies we inhabit.

Modern industrial capitalism based on patterns of high production and high consumption routinely produces high levels of pollution and industrial degradation. International financial and trading systems often frustrate attempts to implement radical change because strict environmental legislation is frequently seen as an impedi-ment to the free movement of capital, labour and commodities. International regimes such as the 1987 Montreal Protocol and possibly the Kyoto Protocol seek to manage (as a technical issue) rather than radically change our industrial and consumptive practices. And perhaps this, above all else, explains why large industrial states such as the US are so reluctant to accept the possibility that our lifestyles are in the long term unsustainable.

Key questions

- How do environmental issues challenge traditional conceptions of geopolitics?
- Why was the 1992 Rio Summit significant in shaping global environmental politics?
- Why are environmental issues intimately linked to North–South relations?
- How can powerful states such as the US and Russia be encouraged to ratify the provisions of the Kyoto Protocol?
- Can global capitalism (including trade and finance) be reformed so that current rates of extraction and pollution are slowed or even reversed?
- Should taxes be levied on aviation fuel, given that commercial aircraft flights generate 600 million tonnes of CO_2 per annum?

Further reading

G. Piel, *Only One World* (New York, United Nations, 1992), J. Palmer, 'Towards a sustainable future' in D. Cooper and J. Palmer (eds.), *The Environment in Question* (London, Routledge, 1992), A. Dobson, *Green Political Thought* (London, Routledge, 1990), G. Prins (ed.), *Threats Without Enemies* (London, Earthscan, 1993), L. Elliot, *The Global Politics of the Environment* (London, Routledge, 1998), G. Porter and J. Brown (eds.), *Global Environmental Politics* (Boulder, Westview, 1996), J. Gupta, *The Climate Change Convention and Developing Countries: From Conflict to Consensus?* (Dordrecht, Kluwer, 1997) and G. Graham, *Ethics and International Relations* (Oxford, Blackwell, 1996). For a critical geopolitical analysis see S. Dalby, *Environmental Security* (Minneapolis, University of Minnesota Press, 2002) and for a feminist critique see J. Seager, *Earth Follies: Coming to Feminist Terms with the Global Environmental Crisis* (New York, Routledge, 1993). M. Paterson, 'Car culture and global environmental politics', *Review of International Studies*, 26 (2000). N. Castree 'The geopolitics of nature' in J. Agnew *et al.* (eds.), *A Companion to Political Geography* (Oxford, Blackwell, 2003).

Websites

United Nations Environment Program www.unep.org
Greenpeace www.greenpeace.org
Friends of the Earth www.foe.org
Basel Convention www.basel.int
Law of the Sea www.un/org/dept/los

Chapter 7

GLOBALIZATION OF HUMANITARIANISM

Key issues

- What are human rights?
- Why is the protection of human rights a matter that potentially transcends the sovereignty of states?
- Why has humanitarian intervention been geographically selective?
- Is the right to humanitarian intervention universal?

Do these places sound familiar? Bunia, Drodro, Kalonge, Bouali and San Carlos de Antioquia. No. Well, all these villages and small towns have experienced appalling atrocities in recent years. In the case of the Congolese village of Bunia, for example, in April 2003 militias using machetes, axes and knives massacred 966 villagers who were tragically in the wrong place at the wrong time. Over the last five years, at least 4 million people have died as war, disease and starvation have taken a terrible toll on the people of the Congo. Since the overthrow of the dictator, Mobutu, the country formerly known as Zaire has been racked by civil war and international intervention. For many Central Africans, the protection of human rights is a chimera. A decade earlier, Rwanda was the scene of a genocide which claimed the lives of at least 800,000 and the International Tribunal based in Tanzania investigating 'crimes against humanity' has thus far handed down guilty verdicts on just nine perpetrators. Given the length of time taken to secure these convictions against members of murderous militias, the victims of the Rwandan genocide may well have to wait for decades if not a century to secure any form of justice (Cockburn and Zarkov 2001).

Why do these atrocities concern us? First, these outrages not only diminish our collective humanity but also demand our empathy given the scale of suffering involved. Second, these kinds of atrocities, whether committed in Central Africa or in European spaces such as Kosovo, sit uneasily with the global vision adopted by the founding states of the United Nations (Fig. 7.1). How can the international community of states allow the flagrant abuse of human rights to stand unchallenged? Is there not a duty (moral and/or legal) to act in response to 'crimes against humanity' regardless of geographical location? As we shall see, the principles attached to state sovereignty and non-intervention often sit uneasily with broader (arguably transboundary) commitments to human rights and humanitarianism.

Figure 7.1 Kosovar Albanians flee from Kosovo in 1999
Photo: PA Photos

The Universal Declaration of Human Rights, adopted by the United Nations
General Assembly in 1948, is a global charter of the rights and responsibilities of individuals and states. Nation-states were considered to be the principal mechanism for
the promotion and protection of human rights. If properly implemented, the UN Charter
would, in principle, undermine a fundamental assumption (in realist and liberal thought)
of international politics – it would seek to ensure that human-rights violations were
addressed regardless of state sovereignty. In other words, repressive governments or
regimes could not, in theory at least, engage in massive human-rights violations and
simply expect the rest of the world to ignore their actions.

There is, therefore, a potential tension (which has arguably been well and truly exposed
in the last decades) between the rights and obligations derived from nation-states and
the human and legal rights endowed by the 'international community' (see In focus 7.1).

Reading these extracts in the aftermath of the 2003 US-led assault on Iraq, it is
striking how Blair's 'Doctrine of the International Community' (a term never defined
by Blair) speech only refers to two individuals – the former leaders of Serbia and Iraq.
Both have been removed from office after US-led assaults on their governments. Blair
does not mention other governments (for example, pro-West regimes such as Saudi
Arabia or Great Powers such as China and Russia) which stand accused of abusing
human rights. The 'Doctrine of the International Community', such as it is, appears
to be selective.

In focus 7.1: The Doctrine of the International Community

In the midst of the NATO assault on Serbian forces in Kosovo, British prime minister Tony Blair delivered an important speech on 22 April 1999 to the Economic Club in Chicago.

With reference to the human-rights violations in Kosovo, Blair argued that 'Awful crimes that we never thought we would see again have reappeared – ethnic cleansing, systematic rape, mass murder . . . [With reference to NATO's justification to bomb Serbian positions] This is a just war, based not on territorial ambitions but on values'.

Linking NATO's operation in Kosovo to globalization, the prime minister contended that, 'Globalization has transformed our economies and our working practices. . . . We live in a world where isolationism has ceased to have a reason to exist. . . . Many of our domestic problems are caused on the other side of the world. Conflict in the Balkans causes more refugees in Germany and here in the US. . . . We are all international-ists now, whether we like it or not. . . . We cannot turn our backs on conflicts and the violations of human rights within other countries if we want to be secure. . . . We need new rules for international co-operation and new ways of organizing our international institutions. Many of our problems have been caused by two dangerous and ruthless men – Saddam Hussein and Slobodan Milosevic. Both have been prepared to wage vicious campaigns against sections of their own community. . . . If we wanted to right every wrong that we see in the modern world then we would do little else than inter-vene in the affairs of other countries. We would not be able to cope.'

Source: www.ndol.org

Geographical selectivity:
to intervene or not to intervene?

For students of critical geopolitics, this accusation of selectivity raises an important question regarding places. Why has the United States and its allies intervened in Iraq (1991 and 2003) and Serbia (1999) and not Rwanda and/or the Congo? Far greater human-rights abuses have occurred in Central Africa. During the Cold War, millions perished either in Russian and Chinese labour camps or via state-sanctioned executions. Why would the international community never tackle large states such as Russia and China over their human-rights records? Human-rights activists believe that the answer lies in the simple fact that these are nuclear-weapons states. The 'international community's' willingness to protect universal human rights over the last fifty years has been patchy, as the human-rights trials currently in progress (in connection with Bosnia, Rwanda, Serbia and Nazi legacies) would also attest.

In the last 50 years, the geography of human-rights protection has tended to favour the wealthier Northern states and their citizens rather than the poorer Southern states, which have (in some cases) experienced brutal regimes and massive human-rights violations. The recent creation of the International Criminal Court (ICC) in The Hague is significant insofar as it signals a determination of 89 ratifying states to address the problem of human-rights violations across continents and cultures. Significantly, however, the United States and small countries in Latin America, Eastern Europe and parts of Africa and Asia have refused to ratify the Treaty of Rome for fear of compromising their sovereign authority. Governments continue to disagree on the best mechanism to protect human rights and the United States in particular fears that its personnel could be subjected to vindictive international legal action.

This chapter concentrates on the position of human rights within the global political agendas of the post-Cold War era due to the growing international profile of human rights and the increasing demands of humanitarian assistance. The first part of this investigation will consider some of the serious political problems relating to the conceptualization and defence of human rights in the absence of universal consensus on the meaning of human rights. The meanings attached to human rights are contested. For post-colonial critics, human rights are considered to be part and parcel of a Western doctrine of rights which is insufficiently sympathetic to the diverse cultures and communities of the world. From that vantage point, human rights can only be culturally specific rather than universally applied.

The second part of this chapter considers humanitarian intervention in world politics and questions whether any intervention can be justified on the basis of human rights (see Vincent 1974). Should poverty and underdevelopment be grounds for non-violent interventions of the kind that some countries and observers demanded for places suffering from genocide, ethnic cleansing and famine? It could be argued that the humanitarian needs of many citizens have been seriously neglected, considering that around 2 million children have been killed since 2000 and that over 2 billion people lack clean and regular drinking water.

The chapter concludes with the problem of how to incorporate these issues into political agendas dominated by states, international organizations and national interests. Human rights and the practices surrounding humanitarian intervention demonstrate that our understandings of global politics need to be broadened.

Conceptualizing human rights: the problem of definition and implementation

Human rights have long occupied a place in Western political thought, from the thirteenth-century British document the Magna Carta to the eighteenth-century Bill of Rights in the USA. This sustained interest in rights, responsibilities and natural law encouraged humanitarian organizations such as the Anti-Slavery Society in the nineteenth century to extend Western conceptions of human rights to non-European peoples. While the spread of humanitarianism is frequently described as a Western phenomenon, it is nonetheless important to recognize that non-Western societies and faiths have demonstrated considerable compassion and responsibility towards the vulnerable, weak and endangered. For the purpose of this chapter, however, attention will concentrate on international law and practice which has evolved since 1945.

Definition box: Human rights and humanitarianism

Human rights are rights held to be justifiably belonging to any person. These rights might include individual rights such as the right to free expression or fair trial and/or social and economic rights such as the right to full employment. However, for others the focus on human rights reflects a Western tradition based on the self-contained individual rather than a more community-based perspective.

Humanitarianism refers to the practice of offering and delivering assistance (often through external parties) for the specific purpose of providing relief from suffering as a consequence of so-called natural disasters such as famine and/or politically motivated disasters such as genocide.

The definition of human rights has been a contested affair. For Western observers, political and legal rights such as freedom of assembly have been conceived in a liberal-democratic tradition which stressed the rights of citizens in relation to the state, although the relative values often varied depending on the interaction of liberal and democratic agendas (Gearty and Tomkins 1996). The extension of rights to women and ethnic minorities was a long drawn-out process even in these liberal-democratic nations. African-American citizens were still struggling for their basic rights even in the 1960s and there was evidence of systematic disenfranchisement in states such as Florida even during the 2000 presidential election (see Chapter 9). For socialist observers, human rights have been conceptualized in broader terms to include social and economic rights such as full employment. The creation of the Soviet Union in 1917 was premised on the belief that social and economic rights would compensate for the loss of formal political rights (see Lane 1996). The International Covenant on Economic, Social and Cultural Rights in 1966 recognized that there were certain social and economic rights and effectively confirmed that the UN Commission on Human Rights had failed to agree on a universal codification of the Declaration. This 1966 Covenant came into force in 1976 at the same time as the International Covenant on Civil and Political Rights. So, even among the Northern industrialized countries, there was (and is) no necessary consensus about the meaning and extent of human rights.

Post-colonial critics and indigenous groups have argued that these definitions of human rights (whether civil, political or economic) ignore the fact that some groups are concerned to preserve their own cultures (in their own terms) rather than be forced to adopt hegemonic understandings of human rights. The most common meaning attached to the term 'post-colonial' refers to the ending of predominantly European colonialism and the emergence of post-colonial states such as the former British colonies in Africa and Asia. In this context, it refers to the removal of external forms of control and exploitation as witnessed during the era of the British Empire. The term 'post-colonial' also applies to ways of thinking about the world, and so-called post-colonial critics have argued that many conceptions of politics, human rights and economic management are based on Western assumptions of individual freedom, liberal democracy and market economies. African and Asian observers believe that there can be few universal human rights because that would imply that the entire world agrees on what actually constitutes human rights (see Walker 1988).

The 1981 African Charter on Human and Peoples' Rights recognized that the demands of African peoples were different from other peoples and implicitly argued that human rights have to refer to more than formal political and legal rights. The right to tribal survival and the preservation of African cultural values was a central theme in this document, unlike the UN Universal Declaration, which focused on the rights of peoples rather than individuals. In a similar vein, the American Convention on Human Rights (1969) recognized specific Latin American concerns over human-rights protection. Accordingly, many African, Latin American, Asian and Pacific observers perceive universal human rights to be little more than an imposition of Western values and belief systems. Arguably this concern has worsened in the aftermath of the United States' public rejection of the ICC, thus exposing perceived American hypocrisy – the public champion of freedom and liberty does not wish to be subject to an institution which may interfere with the domestic jurisdiction of the United States. The Bush administration, for example, fears that American soldiers might be brought before the ICC to face charges relating to their military and humanitarian duties around the world, and that the US will not be able to prevent such action. Powerful states in particular are worried that an evolving global human-rights culture could be used to stifle their freedom to intervene in other places.

Feminist writers have argued that Western understandings of human rights are also rooted in patriarchal assumptions of the role of women and families (see Pettman 1996: 208–11). During the United Nations Decade for Women (1975–85), it became readily apparent that women were systematically disadvantaged in terms of sexual rights, property ownership, legal protection and access to health and education (Enloe 1989, Afshar 1998). In that respect, existing norms and values reflect the experiences of men and hence cannot be universal because they ignore the experiences of women. Moreover, the dominant conceptions of human rights fail to recognize that women are more often in need of protection in the home, where most violence against them occurs, rather than in the public sphere. This argument could, of course, be extended to children, the disabled and the elderly, but the human-rights discourse tends to focus on formal politics and the public sphere.

This exclusion of gender from universalistic conceptions of human rights is further compounded by a lack of recognition regarding social and economic rights. The distribution of global income is highly unequal, with only an estimated 1 per cent of property, land and financial resources being held by women (cited in Bretherton 1996: 256). A commitment to social and economic rights would have to be grounded in an appreciation of the widespread exclusion of women from the ownership of wealth. Changing the rights of women would involve some fairly fundamental reorganization in the world's political economy. It was not until 1984, for example, that the United Nations Commission on Human Rights recognized that domestic violence against women should be a subject for human-rights discussion. Subsequent UN conferences in Vienna (1993) and Beijing (1995) have continued the political debate and policy discussions over the role of women in international politics and humanitarianism (Haynes 1996).

Defining and then defending human rights remains a problematic venture. Some liberals argue that cultural survival and/or environmental security are not really 'rights' in the first place. Alternatively, there has been much criticism from organizations such as Amnesty International that well-established political rights such as those governing torture and illegal imprisonment are frequently overlooked if committed by large powerful states such as China, Russia and the United States. Human-rights lawyers and critical political commentators endorse the notion that legal obligations regarding the defence of human rights often appear to be sacrificed in the *realpolitik* of national interests. Would any politician seriously suggest that China should be forcefully removed from Tibet or Russia physically prevented from abusing the human rights of the Chechen population?

From a feminist perspective, the protection of women's rights also tends to be haphazard, as witnessed in Bosnia, Congo and Rwanda where the mass rape and mutilation of women and female children remains endemic. In 1994, a special rapporteur was finally appointed by the United Nations to highlight 'Violence against Women' and the inadequacies of current protective measures. However, there is also now evidence that such violations against women continue even in the aftermath of UN-sponsored humanitarian intervention. Indeed, humanitarian intervention might actually make the problem worse. By seeking only to provide (temporary) humanitarian relief, outside providers may lack the necessary political mandate or long-term resources.

Enforcing human rights by states: national and regional variations

The United Nations Declaration of Human Rights (1948) remains the landmark document in terms of international legal obligations. Any discussion of human-rights protection and international obligations has to consider this document carefully. By building upon the sentiments of the United Nations Charter and the establishment of the UN Commission on Human Rights, it embodies a series of so-called first-generation rights to political freedoms such as the right to freedom of speech and choice of religious denomination. It was adopted in the United Nations General Assembly by 48 votes to 0.

Its significance lies in the fact that it seeks to define a code of international behaviour while stressing that universal standards should be respected by all member states of the United Nations. It was composed of 30 articles, which covered civil and political rights as well as a range of economic and social rights. Article 1 enshrined the principle that 'all men [sic] are born free and equal in dignity and rights'. Other articles dealt with the freedom to choose a religion, the right to education, and the right to be secure from the threat of torture and illegal imprisonment.

To most Western observers, the United Nations Declaration was not a problematic document as it secured political and legal rights already enjoyed by the majority of these nations. While there was some debate to establish whether the Declaration should seek to protect a narrow range of rights or to be more progressive, most Western nations were in accord with the importance of protecting civil and political rights.

However, there was also significant opposition to the Declaration, which demonstrates the contested nature of human rights. During the vote on the Declaration, eight abstentions (Soviet Union, Ukraine, Byelorussia, Czechoslovakia, Poland, Yugoslavia, Saudi Arabia, South Africa) were recorded. The South African apartheid regime, established in 1948, had effectively excluded the majority of the population from the political system (see Chapter 3) and could not accept the challenge to its domestic authority that the human-rights obligations enshrined in the Declaration would constitute. The Soviet Union and its political allies abstained because it was argued that the Declaration took no account of social and economic rights. The Saudi Arabian monarchy objected because the freedom to choose one's religion (Article 18) violated Saudi law, which outlawed religious denominations other than Islam. As one of the few non-Western nations that were UN members in 1948, Saudi Arabia's objections to the universalism of the Declaration were to be reinforced at a later stage by the newly independent nations of Africa and Asia in the 1950s and 1960s. Saudi Arabia has also never ratified other international covenants such as those concerning civil and political rights (1966) and continues, for example, to engage in public executions.

It was remarkable that the UN Declaration of Human Rights was negotiated at the start of the Cold War. For the Soviet Union, the Declaration was interpreted as an attack on the Soviets and their allies. Soviet leader Joseph Stalin argued that the USA was implicitly represented as the model for other nations to follow in terms of governance and the provision of human rights but that alternative conceptions of human rights were marginalized by the underlying impulse of this Declaration. Unsurprisingly, Soviet leaders tended to raise other issues such as racial discrimination against black people in the United States (and their subsequent struggle for civil rights in the USA) while the latter condemned Russia for violations of civil and political rights. This is important because there has always been accusation and counter-accusation regarding the international community's protection of human rights. The passing of the Declaration did not remove profound areas of disagreement over the meaning of human rights and the means by which they should be protected.

This underlying geopolitical context prevented the introduction of a more binding covenant and restricted the geographical and legal enforcement of human-rights obligations to be carried out by the international community in a uniform and even-handed manner. Furthermore, the worsening relations between the superpowers meant that human-rights protection often assumed a low policy importance compared to strategic and military factors. For example, in many parts of the world such as Latin America, Southern Africa and South East Asia, the US and its allies were prepared to overlook massive human-rights violations by pro-Western military regimes in order to ensure that communism would not flourish in these regions (see In focus 7.2 and Fig. 7.2). The Soviet Union was no better when it came to ruthlessly enforcing its political control on the communist states of Eastern Europe (see Chapter 8).

In focus 7.2: The 'Dirty War' in Argentina

In the 1970s, Argentina was governed by a series of brutal military regimes, which launched a massive and violent campaign against so-called subversives under the label of the 'Dirty War'. The military juntas argued that Argentine national security was being compromised by left-wing revolutionary elements in society whose sporadic guerrilla activities in certain parts of the Republic were cited as an excuse for a violent national security strategy (Nino 1995). Individuals connected to trade unions, professions, the Catholic church, the media and universities were targeted for persecution. Amnesty International estimated that over 10,000 people were executed, tortured and/or simply 'disappeared' during the period between 1976 and 1981. Victims were often dragged off the streets and bundled into cars, which then headed to a network of detention and torture centres. Afterwards, many of the bodies were thrown out of armed forces' planes into the shark-infested South Atlantic.

In 1977, *Las Madres de la Plaza de Mayo*, a campaigning group, was formed by mothers of the missing relatives, who gathered to demonstrate in the central squares of Buenos Aires. Employing the tactics of peaceful resistance, the group's unwavering vigil against the unmitigated brutality eventually forced the military regime to confront the violence of the 'Dirty War'. It was hoped that the fate of the many missing victims would be resolved. To this day, every Thurday afternoon a group of mothers still gathers at the Plaza de Mayo in Buenos Aires holding aloft photographs of the 'missing'.

The current chief prosecutor of the ICC is an Argentine who protested against the 'Dirty War' – Luis Moreno Ocampo. He was also responsible for prosecuting some of those involved in the abduction and murder of so-called subversives under the military rule of 1970s. The Argentine army was still training commandos how to torture as recently as the 1980s and early 1990s.

Universal human rights have also been a mirage for citizens in East Timor, Argentina, Chile and South Africa. The United Nations undertook some limited human-rights and peacekeeping-related work in Cyprus, Korea and the Middle East, but Soviet violations of human rights in Hungary and Czechoslovakia, although condemned, were not actively challenged because the dangerous machinations of the Cold War frequently prevented powerful Western nations from intervening unless abuses occurred in areas of the world considered to be strategically unimportant and/or governed by weak states.

Non-state enforcement of human rights

Increasingly, especially in the aftermath of the Cold War, the most vigorous defence of human rights was often stimulated by human-rights organizations and associated televisual and print-media exposure of violations (see In focus 7.3).

Figure 7.2   The mothers of the disappeared in Buenos Aires, Argentina
Photo: Klaus Dodds

In focus 7.3: Does television coverage make a difference?

It has been contended that television coverage of humanitarian disasters can make a difference when it comes to explaining why some places attract humanitarian intervention while others do not. The so-called 'CNN factor' has been invoked as an impetus for Northern governments to intervene in places such as Somalia (1992), Bosnia (1995) and Kosovo (1999). But the evidence is far from conclusive. Apart from the problem of trying to differentiate between the importance of television versus the perceived geopolitical significance of a region or state, the US-led intervention in Bosnia in August 1995 came after 51 months of televised fighting and claims of genocide against the Bosnian Muslim population. Likewise, the US did not intervene in Somalia for 24 months despite fairly regular television coverage of famine and suffering. These interventions were arguably easier to justify to a sceptical US public because they were frequently cast as a battle against an 'evil man' such as the former Serbian president, Milosevic.

Media reports of crises and genocides in places such as Algeria, Burundi, Nigeria and Sudan have not attracted any substantial international interventions. One reason might be that Northern states such as the US or the UK do not consider these regions in Africa to be sufficiently strategically significant and/or fear that intervention in 'African emergencies' (with all the attendant assumptions that often prevail about those societies and their capacity for pre-modern tribal brutality) will be complex and time consuming. Other than expressions of regret and protest, intervention would not be countenanced. Alternatively, when a Northern state such as the US decides to intervene, as in the case of Kosovo (1999), television images of human suffering actually help to justify a particular policy decision.

This is important because when the Declaration was introduced in 1948, no mention was given to non-state organizations. Nation-states were assumed to be both the source and guardians of human rights. It is precisely because states have been so variable in their implementation and protection of human rights that, in the midst of the Cold War, the pressure group Amnesty International was founded in the UK in order to monitor human-rights abuses and violations around the world. From 1961 onwards, the London-based organization has campaigned on behalf of those illegally imprisoned, tortured and/or denied basic human justice. Amnesty International relies on voluntary donations and private subscriptions to fund campaigns or specific cases such as the imprisonment of Nelson Mandela in South Africa, the enduring Indonesian violence in East Timor and human-rights abuses in Burma, China and Argentina (Fig. 7.3).

Figure 7.3 Amnesty International poster
Source: Amnesty International Publications www.amnesty.org

It is also important to note that Amnesty's annual evaluations of human rights have in the past included criticism of governments such as the UK government for repressive security policies in Northern Ireland which denied basic civil and legal rights to 'terrorist suspects'. As well as being criticized for their failure to protect the rights of domestic citizens, states such as the UK, the USA and France have been goaded into taking action against human rights violators by the efforts of pressure groups like Amnesty International. There have nonetheless been many examples of Western governments refusing to intervene on behalf of oppressed peoples for commercial and geopolitical reasons, and it has to be acknowledged that enduring problems exist regarding the defence of so-called universal human rights.

Can human rights ever be universal?

It should already be apparent that the idea of universal human rights is conceptually controversial and politically problematic. For supporters of universality, human rights are derived on the basis of a moral argument regarding the intrinsic and equal worth of human beings. Article 1 of the UN Universal Declaration reflects this particular philosophical position by acknowledging that 'all human beings are born free and equal in dignity and rights'. However, the underlying concept of universal rights has been challenged on the one hand by Western critics who take issue with the assumption that human nature is based on the capacity for moral reasoning and rational action, and on the other by post-colonial critics who have questioned Western philosophical assumptions of individual rights. Feminist critics, meanwhile, have censured the gendered assumptions of human-rights discourse and practice.

On a more optimistic note, though, some observers such as Francis Fukuyama (see Chapter 1) believe that with the emergence of liberal democratic governments in the 1980s and 1990s in Latin America, Eastern Europe, Africa and Asia, the spread of democracy enables a greater number of countries to share a particular moral and political consensus. Many people in Central and Eastern Europe greatly value their newfound civil and political rights, but the problem with this kind of argument is that it either ignores the fact that regions such as the Middle East have shown little inclination towards Western models of democracy or neglects the fact that many Asian nations have consistently rejected (Western) civil and political conceptions of human rights. The 1993 World Conference on Human Rights, for instance, witnessed Asian and particularly Chinese opposition to universal concepts of rights.

One way forward for the human-rights agenda might be to concentrate on identifying human wrongs rather than rights. It has been proposed that international law should seek to develop a new code of human wrongs, which could be then used to delegitimize certain actions. Supporters of this strategy have pointed to the 1948 Genocide Convention to illustrate the successful use of international law for the prosecution of officials and soldiers responsible for the Rwandan massacres in 1994–5. This convention, based on the experiences of the 1945–6 Nuremberg War Crimes Trials, introduced into international law the concept of crime against humanity. The prosecution of German military and civilian leaders relied on the assumption that universal and inalienable rights were grossly violated during the Second World War, thus creating a new legal precedent. However, ensuring compliance with laws prohibiting crimes against humanity has been problematic, as the massacre of over 3 million people in places such as East Timor, Cambodia and Rwanda would testify. It is also proving difficult for the international community to prosecute those suspected of massive human-rights abuses even if provisions for the International Criminal Court are entering into force (see Nino 1995 for an earlier review).

The alternative proposed by some philosophers seeks the promotion of non-foundational human rights to overcome the universal impasse. Grounded in an appreciation that cultural relativism can be and has been used by repressive regimes to justify massive human-rights violations, this approach would seek to match cultural traditions with acceptable forms of human rights. At the 1993 UN Conference on Human Rights, India and China condemned universal human rights for being little more than an extension of European understandings of rights and human freedom. The challenge for Western nations determined to uphold a tradition of universal human rights would be to demonstrate that political and social life is preferable in a context where human rights are respected as opposed to violated. International standards on political rights regarding issues such as torture and genocide would have to be protected in tandem with a commitment to respect particular cultural variations. No Chinese or Soviet leader, for example, has ever claimed the *de jure* right to torture political suspects and/or imprison dissidents without trial.

In the post-Cold War era, the protection of human rights remains precarious. Upholding universal human rights is unlikely to be achieved without some consideration of the material conditions of life. Meanings of needs, justice and ownership within different societies would have to be recognized within discourses on rights and democracy (see Baylis and Smith 1997). When the former British foreign secretary Robin Cook met the prime minister of Malaysia, in August 1997, newspaper reporting of the meeting indicated that the long-serving Malaysian premier, Dr Mahathir Mohamad, was unhappy with Cook's rigid adherence to the United Nations Declaration in their discussion of the human-rights situation in South East Asia. The former argued that the Declaration was scripted by and for the rich North and therefore failed to address Asian and African values. Malaysia later expressed reservations about British proposals for further sanctions against Nigeria (in the wake of the Ken Siro Wiwa hanging in 1995) as presented to the Commonwealth Summit in October 1997. Dr Mahathir has declared time and again that he does not believe authoritarianism to be any better or worse than democracy and that sanctions directed against non-democratic regimes accused of human-rights abuses merely tend to hurt the poor of that country.

Recent interventions in Kosovo (1999), Afghanistan (2001) and Iraq (2003) by US-led forces have also created a culture of extreme scepticism towards human-rights protection and humanitarian intervention (Fig. 7.4). Without explicit United Nations approval, US and UK forces (in the main) launched bombing raids on Serbian and Iraqi forces respectively. The bombs were often dropped from over 15,000 feet in order

Figure 7.4   US soldiers in Iraq: are they liberators or oppressors?
Photo: PA Photos

to avoid possible anti-aircraft fire and thus were at times inaccurate. While few would deny that the regimes were governed by objectionable and violent individuals who brought suffering to their own populace, many observers have noted how 'humanitarianism' has repeatedly been used to justify outside intervention.

The danger of such a strategy is threefold:

1. It becomes a convenient means of legitimating action against a regime that certain Great Powers do not recognize and/or endorse.
2. It can be used to divert attention away from the manner in which some states are selected for 'humanitarian intervention' and others are not.
3. In their haste to be seen reacting to a particular crisis, the international community may neglect to consult with local communities who may be actually very anxious about the long-term consequences of outside intervention (whatever the motivation).

Decisions about when and where to intervene are clearly not made on the exclusive basis of human suffering.

At the start of the twenty-first century, then, no global consensus on human rights exists. Although human rights monitoring has improved through the appointment of a High Commissioner for Human Rights (1994), human-rights abuses have been reported in 150 countries (see Amnesty International 2002), so there remains much to do in terms of ensuring basic human-rights compliance. Other UN-based organizations

In focus 7.4: Humanitarian or sexual intervention?

Since October 1999, 16,000 troops have been attached to the United Nations peace-keeping mission (UNAMSIL) in Sierra Leone, West Africa. For much of this period, local women have accused UNAMSIL of either failing to prevent rape by local militias or more disturbingly being complicit in mass rape. Despite a UN Code of Conduct explicitly prohibiting sexual abuse or exploitation, it seems to be a widespread phenomenon in Liberia, Sierra Leone and other parts of West Africa.

This problem of sexual abuse by UN soldiers and support staff is not unique to West Africa, however. In the post-1995 period, a large sex industry has been established in Bosnia following the deployment of 50,000 peacekeepers. Local and international reports have alleged that the creation of brothels throughout Bosnia has encouraged illegal trafficking of young women and girls from other countries, in particular Albania.

The United Nations has no authority to punish offenders and instead relies on contributing states to punish those found guilty of rape and/or sexual contact with underage girls. The tragic irony of peacekeeping operations, therefore, is of women and children being doubly victimized by the very people who are supposed to be improving their lives. 'Security' and 'humanitarian intervention' are never gender-neutral.

such as the Committee on the Discrimination against Women (CEDAW) have attempted to address the problem of women's rights, but underfunding has hampered progress in recent years (see In focus 7.4). This is absolutely critical – men and women often enjoy differential human rights. Despite the Declaration being gender-neutral in tone, women are often subjected to restrictions and violations that clearly compromise their 'right' to seek employment, enjoy a fair trial and/or assume access to community rights. Perversely, UN-sponsored humanitarian intervention may not actually improve the condition of women within affected communities (Enloe 1989, Peterson and Runyan 1993).

Reconceptualizing human rights is only part of the problem and future progress will depend in part upon the development of effective machinery to protect and monitor human rights.

Humanitarian intervention in the post-Cold War era

The capacity of televisual coverage of human suffering to influence or even determine government decision-making has been one of the most commented-upon features of the post-Cold War era (see, for example, Gowing 1994, Robinson 1999). In the wake of a number of high-profile humanitarian missions in Europe, the Middle East and Africa media scholars often refer to this phenomenon as the 'CNN factor' (see Chapter 4). For some observers, the power of television to influence government decision-making is perceived to be a positive feature of global politics because it can mobilize the international community into taking action in response to massive human-rights violations. Yet, political leaders and their advisors have cautioned that television coverage can be selective in the sense that the media concentrates on some places at the expense of others (such as Congo, Kashmir and Angola). In some cases, newspaper reporting neglects Third World human-rights issues in favour of 'European' events such as those in Kosovo, Northern Ireland and Chechnya. Finally, some organizations responsible for humanitarian aid such as the International Committee of the Red Cross have complained that media coverage of human suffering can be too simplistic, and often unwittingly conceals the global processes which perpetuate poverty, hunger and human-rights violations. Media reporting of the Rwandan massacre from April 1994 onwards, for instance, concentrated on ethnic and tribal divisions within Rwanda but failed to explore the role Great Powers such as France had played in supporting a brutal and murderous regime for over 20 years (see Prunier 1995).

Humanitarian intervention has, therefore, emerged as one of the leading areas for debate in discussions on global politics and human rights (see Agnew 1992, 2002 and O'Loughlin 1992). Discussion on the role and purpose of humanitarian assistance has extended the limited realist view of political life, which is either sceptical of international co-operation and/or critical of global aspirations. The provision of relief for distressed peoples and the protection of basic human rights have emerged as central themes in these new humanitarian debates (Pugh 1996, Weiss and Collins 1996). The former secretary general of the United Nations, Dr Boutros Boutros-Ghali, published *An Agenda for Peace*, which placed human rights and peacekeeping at the top of the UN's agenda and demanded that member states commit themselves to the funding of HR-related projects (Boutros-Ghali 1992). He was also a firm supporter of Chapter VII of the UN Charter, which justifies outside interference in the affairs of a state which is guilty of massive HR abuses. In April 1991, for example, Dr Boutros-Ghali supported Resolution 688, which authorized Western jets to protect the airspace above northern Iraq when fears surfaced that the Kurdish people were vulnerable to attack from Saddam Hussein's airforce.

At the heart of these examples lies a geopolitical paradox: how can humanitarian intervention be justified when it occurs within an international political system premised on state sovereignty and norms of non-intervention? For the last three hundred years, a Westphalian society of nations has been based on the assumption that states are responsible for their own affairs and that by implication other states cannot intervene in the domestic affairs of their neighbours. In the aftermath of the Nazi Holocaust and the creation of the United Nations, a new wave of human-rights charters and conventions changed the relationship between nation-states and international society (A. Smith 1995). Intervention could be justified in terms of either self-defence or an attempt to prevent murderous states committing massive human-rights violations. Since the ending of the Cold War, these ideas have been expanded to include consideration of non-military and military forms of intervention by states and non-states, but because of the concerns expressed by Asian and African states that humanitarian intervention could be used as an excuse for great-power involvement in the affairs of the weaker nations these discussions have been confined to the West.

Humanitarian intervention: for and against

The dilemmas concerning humanitarian intervention (HI) touch upon some of the critical issues facing contemporary world politics. In the academic debate, the apparent moral imperative to relieve the suffering of other people has been weighed against the legality or otherwise of intervention (see Proctor and Smith 2000, N. Smith 2001). The Commission on Global Governance has argued, for instance, that intervention is justified in cases where massive violations of human rights have occurred and an international response is required to prevent a further loss of life. The justification of outside intervention and the role of force in order to secure humanitarian objectives have been vexing problems. Some of the issues regarding the use of military force for the purpose of humanitarian intervention are addressed using examples from recent UN involvement in Somalia (1992–3), Bosnia (1993–5), Rwanda (1994–5), East Timor (1998–9) and the controversial intervention affecting Kosovo (1999). The US-led assault on Iraq in 2003 is discussed in more detail in Chapter 9.

The use of military force to support HI has been called into question after the televised failures of the Somalian and Rwandan humanitarian operations. The Americans were widely seen to be the principal supporters of these interventions, but US troops dispatched to the Horn of Africa in December 1992 quickly became embroiled in street fighting with local factions and several so-called warlords. In October 1993, former President Clinton ordered the retreat of the US military from Somalia after the death of 18 soldiers in an armed confrontation with a local militia. Televised pictures of dead US servicemen being dragged through the streets of Mogadishu prompted much internal debate in the USA as to the effectiveness of HI and called for a reappraisal of the UN relief operation. These and other deaths (such as of the hundreds of Somalis who were killed during US military operations), are among the strong reasons for a critical examination of the intervention.

Arguments against intervention

1. Article 2 of the United Nations Charter enshrines the principle of non-intervention. This means that states must respect the sovereignty of other states and refrain from intervention in their affairs. HI is an act which seeks to intervene in the domestic affairs of another state. At the very least military-based HI would be illegal under the UN Charter unless one could demonstrate that wider human security was at risk.

In 1978, Tanzania invaded Uganda with the purpose of overthrowing the murderous regime of Idi Amin. During a period of eight years over 300,000 people had been murdered. Tanzania argued that these activities threatened the security of her own people because the Ugandan security forces had crossed the border in search of further victims. This intervention, whilst not expressed in humanitarian outcomes, had a humanitarian benefit in terms of removing Amin from power. He was later given refuge by Saudi Arabia, and died there in 2003.

2. It has been argued that states sometimes act for ulterior motives. It has been alleged, for instance, that France proposed HI in Rwanda during the 1994 massacre of Tutsis

in order to maintain her influence in Francophone Africa rather than to relieve the human suffering. There is no question that HI would have had the effect of raising France's profile in a region where American influence was growing at the expense of the French. Critics maintain that many large states are not sincerely committed to HI. The 2002–3 American budget for the development agency USAID, for instance, has been reduced and US foreign aid donations have dropped to 0.15 per cent of GDP.

It has also been suggested that the NATO-led intervention in Kosovo was motivated not only by a desire to relieve the suffering of the Albanian-speaking population within the autonomous province but also by a desire to remove the then Serbian government. Without explicit UN authorization, NATO planes began dropping bombs in April 1999 on Serbian military positions in Kosovo and more controversially on the civilian infrastructure of the Serbian capital city, Belgrade. Critics contend that HI was used by the United States to selectively remove so-called 'rogue regimes' (as defined by the US) rather than selflessly promote the global humanitarian condition.

3. The principles of HI have been and will continue to be inconsistently applied by the international community. Sceptics point to the contrasting experiences of the Bosnian Muslims and the Iraqi Kurds. The latter were judged to be in need of HI and in 1991 the American-led operation established a safe-haven policy in Northern Iraq for the purpose of protecting the Kurds and the so-called Marsh Arabs in the South from Saddam Hussein's regime. In contrast, the international community was unprepared to intervene actively in the defence of the Bosnian Muslims in 1992–4 (see Gow 1997). Intervention was only forthcoming after the Srebrenica massacre in July 1995 when thousands of Bosnian Muslim men and children were massacred by Bosnian Serbian forces (O Tuathail 1999). Tragically, the town had been monitored by a small group of poorly armed Dutch peacekeepers.

4. Definitions of humanitarian acts such as seeking to prevent the suffering of others can be problematic in the sense that our understandings change over time and space. Public attitudes in Britain towards slavery, for example, have changed from benign acceptance to outright rejection over the last two hundred years. Understandings of human suffering can also be culturally specific.

5. Defining common principles of HI will always be difficult because they rely on the international community's agreement to place individual and communal justice above the principle of sovereignty and non-intervention. In the absence of any common principles it might be better to avoid legitimizing HI within the canons of international law (see Mullerson 1996, Pugh 1996).

For sceptics about HI, there are good reasons and ample evidence for resisting further international endorsement of such practices. The suspicion remains, throughout many parts of the world, that powerful states such as the United States, China and Russia will always use 'humanitarian intervention' in a highly selective manner. NWS such as India and Pakistan, for example, have never been subjected to HI, despite the high levels of death in the disputed province of Kashmir.

In an alternative vein, a considerable legal and moral case can be brought to bear in favour of HI.

Arguments for intervention

1. The UN Charter commits states to protecting fundamental human rights and preventing abuses. In that context, HI could be seen as a legitimate mechanism for the preservation and protection of human rights. Recent UN resolutions such as 43/101 and 45/100 have enshrined the right of NGOs to assist in the provision of aid to troubled areas and demanded that so-called corridors of tranquillity be created in order to assist the delivery of aid. NGOs such as the International Committee of the Red Cross have emerged as important sites of advice and consultation.

2. In a moral context, it has been argued that the international community has an obligation to ensure that states respect basic human rights. Under the canon of international law dealing with genocide and human rights, HI would be a morally and legally legitimate form of intervention if evidence of massive oppression and violation existed.

3. There is evidence to suggest that HI is part of customary international law and that states have recognized in the past that intervention might be justifiable. The Vietnam invasion of Cambodia in 1979, whilst not framed in humanitarian terms, demonstrated that the international community will tolerate certain actions if they are seen to result in humanitarian benefits such as the ending of a genocide by murderous figures such as the late Pol Pot.

4. With the termination of the Cold War, there is a real possibility for the international community to develop a new moral and legal consensus on norms of behaviour. The UN's budget for peacekeeping and humanitarian projects has grown substantially and in 2002–3 the UN was involved in 13 operations with the assistance of 40,000 personnel. Ten years earlier, at the height of the Bosnian crisis, the UN had 70,000 personnel involved in 17 operations.

Clarifying the legal and moral position of HI is problematic given the lack of consensus. For supporters of HI, the international community should make every effort to develop a consensus around which the principles of HI could be firmly elucidated. For much of the Cold War, the international community rarely considered HI because strategic rather than humanitarian intervention tended to dominate international affairs. The 1990s presented new opportunities to consider how humanitarian outcomes could be secured through intervention. For sceptics, however, HI if ever formally legitimated by the 'international community' could undermine the contemporary international order based on the principles of sovereignty and non-intervention. Realists suggest that the 'international community' should not seek to develop new norms of governing HI because of their potential to undermine the existing international system.

Contemporary humanitarian intervention: enduring tensions

In the contemporary era, Western sceptics of HI have in some sense been overtaken by events. Televised reporting of human suffering has had a powerful influence on domestic public opinion in regions such as Europe and North America. For better or for worse, political leaders continue to confront the arguments for and against HI in an unprecedented fashion. Television coverage can be used to force political leaders to commit themselves to intervene in order to relieve suffering. Alternatively, media reporting of humanitarian disasters can also make HI politically unattractive in places such as the USA. Among non-Western nations, however, the current propensity for HI discussions in the West is regarded with some suspicion. China has been consistently sceptical of Western motivations for intervention and as a veto-carrying member of the Security Council of the UN is likely to be hostile to granting HI a legitimate status.

There are a number of issues worth exploring in some detail regarding HI. The first concerns the legitimacy of HI and the mechanisms used in the UN to gain approval for such actions. The chances for success or failure of HI have to be established because this will have direct implications for the future legitimacy of HI. The use of force is considered in the deployment of HI because in the recent past it has been necessary to protect the operations of the UN agencies. The role of NGOs such as the Red Cross and Médecins sans Frontières can play in the delivery of HI should also be considered because it is clear that states (contrary to realist assumptions) have to co-exist with other non-state parties in the international humanitarian arena (Fig. 7.5).

It has been noted that in the post-Cold War era the UN Security Council has rarely legitimated HI on the basis of humanitarian grounds alone. Why is that? Whilst

Figure 7.5 Médecins sans Frontières (Doctors without Borders) is an extremely important NGO which works in close co-operation with states and the United Nations
Photo: PA Photos

humanitarian considerations were undoubtedly important in motivating UN operations in Somalia and Kurdistan, Western political leaders frequently drew upon other factors to justify the relief of massive human suffering. In the case of Kurdistan, for example, UN resolutions (such as 688) governing military-backed humanitarian intervention were justified by Western leaders on the basis of 'protecting' the international peace and security of the region in the light of Saddam Hussein's repression in Northern Iraq. Other observers such as China objected to the approval of military enforcement action (such as the declaration of a no-fly zone in Northern Iraq) because of concern that this action would establish a precedent for use of military force in the defence of human rights. The refusal to endorse a military enforcement mandate by China and the Soviet Union meant that the US and its allies used Resolution 688 to employ other parts of the UN Charter to justify military intervention.

In contrast, Resolution 794 approved the deployment of American forces and their relief operations in Somalia. Military intervention was justified and approved by the Security Council because the Somali state had collapsed and violent civil war had broken out. Moreover, in contrast to the Kurdistan operation, there was no direct opposition to the intervention from a de facto leader of state. The norms of non-intervention and state sovereignty had not been undermined by this US-led intervention. As a failed state, Somalia was not even considered by sceptics such as China to present a general precedent for HI and the final draft of Resolution 794 contained numerous references to the unique and unusual nature of the Somali case.

The effectiveness of the UN Somalia operation was modest. In the short term, there can be no doubt that the operation led by the US did relieve the suffering of starving civilians. In the longer term, however, the plan to demilitarize Somali society was called into question because the UN's and US's apparently neutral objective of providing humanitarian aid appeared to be replaced by a desire to capture particular warlords for the purpose of imposing order. Television coverage of US helicopters hovering over the streets of Mogadishu appeared to confirm that this humanitarian operation had been transformed into a military exercise, with disastrous results. As a consequence, the local population lost trust in the Americans and came to believe that the intervention was not motivated by humanitarian impulses.

Case study: the United Nations and Yugoslavia (1992–5)

The United Nations response to the crisis in the former Yugoslavia (1992–5) brought to the fore the complex interrelationship between moral claims regarding human rights and norms regarding state sovereignty and non-intervention (Woodward 1995, Weiss and Collins 1996 and Fig. 7.6).

This example of humanitarian intervention had substantial consequences for post-Cold War global and European geopolitics. The crisis in the Balkans was the first example of the UN operating in the heart of Europe as opposed to the so-called Third World, and billions of dollars were invested in conflict and refugee management.

Figure 7.6   The Yugoslav towns of Zagreb and Dubrovnik were popular holiday destinations for Western tourists before the conflict in the early 1990s
Photo: Klaus Dodds

However, the apparent failure of the UN to intervene decisively in 1992–3 in the face of human-rights violations and ethnic cleansing caused a major crisis of confidence in HI provision. The UN did not attempt to stop these violations; rather it approved a form of peacekeeping which attempted to provide humanitarian assistance without active engagement with the parties responsible for genocide and the destruction of multi-ethnic communities. Former UN commanders such as General Francis Briquemont frequently complained that UN Security Council resolutions regarding the former Yugoslavia were not being enforced in the field because of a lack of political will to enforce peace accords. The UN peacekeeping force in Croatia was only given a $250 million budget, when experts had estimated that around $700 million was required to fulfil the UN peacekeeping and humanitarian objectives (Weiss and Collins 1996).

The deployment of 14,000 UN peacekeepers in 1992 to the Republic of Croatia was not able to prevent the spread of war in the former Yugoslavia or to empower Croatian refugees to return to former villages and towns. The imposition of no-fly zones in Bosnia in 1993–4 was rarely enforced and the so-called protection of safe havens was piecemeal and haphazard. Over 40 Security Council resolutions urging the parties to call an end to the fighting in the former Yugoslavia were passed with little impact (see Mullerson 1996). HI was justified in 1992–4, on the basis of threats to international peace and security, but was largely ineffective because relief convoys were not adequately protected from Serbian and Croatian armed factions. By late 1993, 80,000 UN peacekeepers had been deployed in the Balkans for the purpose of humanitarian assistance and conflict resolution, and in spite of the massive increase in peacekeepers, the debt-ridden UN proved incapable of enforcing peace in the region.

In February 1993 it was mooted that the UN, because of the persistent failure of warring parties to accept agreements over access to aid for civilians, should end calls for HI. Some commentators were already calling into question the financial and political wisdom of HI in the face of other resource claims regarding poverty and preventable diseases. Others demanded that the UN create a new humanitarian delivery unit with clear rules and procedures for the delivery of HI (Weiss and Collins 1996). Inadequate military and humanitarian action in the former Yugoslavia was compounded by doubts amongst Western allies as to the effectiveness of air strikes, sanctions and on-the-ground military intervention. Media reporting of human suffering combined with the 'something must be done factor' cast considerable doubt on the effectiveness of HI.

Within the United States, considerable soul-searching ensued regarding the provision of HI and the use of military force in the wake of the Yugoslav crisis. The experience in Somalia led conservative commentators such as Senator Sam Nunn to argue that the US should not commit troops to Bosnia for the purpose of conflict resolution. In other words, it was not in the national interest of the United States to intervene to relieve the suffering of others. This view was effectively overturned in 1995 when the US helped broker the 1995 Dayton Accord. This finally secured a modicum of peace and territorial stability in the region even if it actually confirmed the ethnic-cleansing gains of the powerful Serbian and Croatian factions. Fifty thousand NATO troops, deployed in a peacekeeping role, helped to ratify the status quo ante. Bosnian Muslims complained that this form of intervention simply failed to address territorial grievances and the illegal actions of aggressive factions and neighbouring states. Peace may have been secured but at the expense of any kind of justice.

In terms, therefore, of humanitarian outcomes (as opposed to motives), HI in the former Yugoslavia achieved mixed results. Whilst it brought some relief to the suffering of civilians in the war zones it failed to prevent the killing of Bosnian Muslims by Serbian forces in towns such as Srebrenica (Honig and Both 1995, Simms 2001). The toll of the Yugoslav wars was, however, dreadful in the sense that an estimated 200,000 died, 2.7 million people were made homeless and 6 million anti-personnel land mines remain buried in the region.

Geographies of intervention and non-intervention

China and India have argued that state sovereignty does not permit outside intervention even though powerful states such as the USA and the UK appear to use the UN Charter's definition of 'international peace and security' to suit their own agendas. Selective application of such a definition means that some countries such as Somalia are deemed worthy of HI whilst neighbours such as war-torn Sudan (until recently) are ignored. Some humanitarian workers have reached similar conclusions and have argued that HI can often be counter-productive and strategically selective (see Weiss and Collins 1996). Whilst the US and her allies through the Security Council have secured the *droit d'ingérence*, Western-backed HI does not appear to have been balanced by a commitment to long-term conflict resolution and redevelopment. These concerns were recently raised again following US operations against Kosovo (1999), Afghanistan (2001) and Iraq (2003), which had been justified at least partially in humanitarian terms (see In focus 7.5).

In the longer term, the role and scope of HI will have to be clarified at the very least in terms of the deployment and development of the humanitarian roles of the armed forces, the influence of television coverage and the role of non-state actors (see P. Taylor 1997). As far as sceptics are concerned, the militarization of HI provokes further levels of violence in crisis-ridden regions and complicates the process of reconciliation and rebuilding, while proponents argue that it may be necessary in the face of rampant human-rights abuses and genocidal violence. In the absence of an effective government, as in the case of Somalia, the provision of militarized HI could be justified in terms of seeking to reduce suffering in the face of armed militias imperilling citizens. As the UN secretary-general Kofi Annan noted in 1993, 'The reality is there are situations when you cannot assist people unless you are prepared to take certain military measures' (cited in Weiss 1994: 6).

The role of non-state organizations such as the Red Cross also raises issues concerning the management of HI. Trans-national networks of humanitarianism are changing the remits of provision and organization and make the promotion of non-statist and non-military forms of HI a pressing challenge for both Western and non-Western critics. The ability of NGOs and social movements to create a new moral

In focus 7.5: A fig leaf for Great Power politics? Humanitarian intervention and NATO's operations in Kosovo (1999)

The distinguished historian Eric Hobsbawm, commenting on the NATO assault on Serbia and Kosovo, noted that wars are rarely fought for exclusively humanitarian reasons. NATO's assault in March–April 1999 was motivated for reasons that were far from humanitarian. According to critics such as Tariq Ali, Peter Gowan and Edward Said, the assault was motivated by a desire to prove the US and NATO's credibility in the post-Cold War era. The United States, so Defense Planning Guidance produced by the Pentagon seemed to imply in 1992–3, was determined to prevent others such as Russia from developing a more assertive regional let alone global role. Moreover, the US was also anxious to maintain NATO as a US-led security arrangement rather than encourage a European-led security union such as the OSCE to intervene in the Kosovo crisis. Significantly, this was the first time NATO forces had taken offensive action since the creation of NATO in 1949.

As a military operation it was far from successful, however. After 78 days of high-level bombardment, Serbia's military machine was largely unscathed while the civilian infrastructure of Belgrade was badly destroyed. Although the despised Serbian leader Milosevic was eventually ousted and brought before an International Tribunal in The Hague, the bombing of Kosovo arguably worsened the ongoing refugee crisis. Kosovar Albanians were forced to flee in greater numbers from the autonomous province for fear of being hit by high-level NATO bombing. NATO forces remain in Kosovo and the minority Serbian community has been forced to flee as Albanian-speaking inhabitants dominate the interim administration. Critics, especially in the aftermath of US operations in Afghanistan and Iraq, remain convinced that the assault on Serbia was motivated by a desire by America to exert its dominance in an uncertain world.

Further reading: *see* Ali 2000.

consensus on humanitarian assistance should not be overrated, as the globalization of humanitarianism is still a long way off, judging by the lack of intervention in places such as Rwanda, Angola and Liberia and the opposition of countries such as China to HI (Pugh 1996, Prunier 1995, Shaw 1996).

While prosecuting the perpetrators should become easier in the aftermath of the establishment of the ICC, there are still substantial obstacles to confront. Powerful states continue to seek further 'protection' from international bodies charged with human-rights protection. The refusal of the United States to ratify the Treaty of Rome has provoked the Bush administration to seek so-called 'immunity deals' with separate countries which would ensure that American personnel in these countries would never be referred to the ICC. Over 50 countries have thus far been persuaded to offer the US 'opt out' deals. Moreover, as a permanent member of the Security Council, the US could block any referrals to the ICC and these might include the prosecution of dictators in places where the Treaty of Rome has not been ratified (e.g. Iraq). Finally, no crime committed before 1 July 2002 can be addressed by the ICC – hence the creation of ad hoc Tribunals for Rwanda and Bosnia.

Conclusions

Since the Second World War, there has been a considerable extension of international law regarding moral standards of governance. The Universal Declaration of Human Rights remains highly significant in terms of establishing the specific responsibilities of government *vis-à-vis* the citizen. More recently, the possibility of universal human rights has been advanced by some critical commentators to suggest that a global civil society could emerge to establish new sets of relationships between the individual, the state and the world. The creation of the ICC must be seen as an important step towards this goal. In the post-Cold War era, the norms of non-intervention and state sovereignty have been challenged by events in Europe and beyond. Old debates about national interests and self-interested intervention are required to co-exist with new agendas based on collective security, common humanity and human-rights protection. This does not mean, however, that all states (as part of an international community) agree on how these elements should be combined with one another. At present, despite some claims to the contrary, the role of states will remain critically important in shaping this national/global human-rights equation.

Debates over humanitarian intervention and human rights will continue throughout the twenty-first century. Any criteria for judging the legitimacy of intervention will have to consider the following: the nature of the authority approving the intervention, the motivation for the intervention and the outcome of the intervention. Difficult issues will have to be confronted, such as how one judges the nature and extent of the suffering which might justify intervention in cases where there is no proven record of massive human-rights violations. Televisual coverage of human suffering does not necessarily improve the capacity of decision makers to adjudicate on the nature of particular violations. Media coverage of human-rights violations can be partial and/or ignore events in places such as Saudi Arabia, Congo and Kashmir (see Chapter 4). The geographies of intervention remain controversial. In the future, the role of the military in meeting HI needs to be carefully assessed because the experiences of UN-sanctioned and other non-sanctioned operations in West Africa, South East Europe and even Iraq demonstrated that military-based intervention could have a disastrous impact on local, often vulnerable, societies. US soldiers in Iraq, for example, would be the first to admit that they have not been welcomed with open arms following the overthrow of the Saddam Hussein regime.

Human rights and HI are therefore intimately connected to the political and economic globalization of the planet. The provision of humanitarian aid illustrates a growing trend in global politics towards intervention in the affairs of other states. The longer-term challenge for proponents of HI is not only to establish clearer criteria for this employment but also to recognize that crises in places such as Bosnia, Kosovo, Congo and Somalia are products of a highly unequal global system which contributes to unstable local and regional political conditions. Global solutions to humanitarianism will depend on adequately addressing issues relating to warfare, poverty, maldevelopment and despair.

Key questions

- Why should geopolitics consider the issue of human rights and humanitarian intervention?
- Does television coverage of human-rights abuses make a difference?
- Has the protection of human rights become more important in the post-Cold War era?
- Why was there no humanitarian intervention during the 1994 genocide in Rwanda?
- What considerations shaped European and American responses to the Bosnian Crisis (1992–5)?
- Is humanitarian intervention the charitable front of contemporary neo-imperialism?

Further reading

On human rights see D. Byrne, *Human Rights* (Harlow, Pearson Education, 2004) and C. Gearty and A. Tomkins (eds.), *Understanding Human Rights* (London, Mansell, 1996). On some of the debates surrounding humanitarian intervention see O. Ramsbotham and T. Woodhouse (eds.), *Humanitarian Intervention* (Cambridge, Polity, 1996), A. Roberts, 'Humanitarian war: military intervention and human rights', *International Affairs* 69 (1993): 429–49, T. Weiss, 'UN responses in the former Yugoslavia: moral and operational choices', *Ethics and International Affairs* 8 (1994): 1–22, and T. Weiss and L. Collins, *Humanitarian Intervention in the Post-Cold War Era* (Boulder, Lynne Rienner, 1996). On gender and human rights see C. Enloe, *Bananas, Beaches and Bases: Making Feminist Sense of International Relations* (Berkeley, University of California Press, 1989) and V. Peterson and A. Runyan, *Global Gender Issues* (Boulder, Westview, 1993). A useful review of the role of television is P. Robinson, 'The CNN effect: can the news media drive foreign policy?', *Review of International Studies* 25 (1999): 301–9. A classic, if older study of intervention in international society, is R. J. Vincent, *Non-Intervention and International Order* (London, Routledge Kegan and Paul, 1974).

Websites

Amnesty International www.amnesty.org.uk
Human Rights Watch www.hrw.org
Médecins sans Frontières www.msf.org
United Nations www.un.org

Chapter 8

ANTI-GEOPOLITICS AND GLOBALIZATION OF DISSENT

Key issues

- What is anti-geopolitics?
- How were colonialism and imperialism represented and resisted by Third World writers such as Edward Said?
- How was dissent expressed during the Cold War in the former communist states of Eastern Europe?
- Is anti-globalization a misnomer?

For much of the twentieth century, geopolitics has been synonymous with global conflict and change, as states struggled for mastery of power and space. Unsurprisingly, therefore, the intellectual history and political practices associated with geopolitics tend to reflect the conceits and interests of those who enjoy positions of political, economic and cultural influence and power. Issues pertaining to nation-states and statescraft have prevailed, often at the expense of those who contested and/or resisted the power of the state and geopolitical practices. As with the official chroniclers or producers of History, our understandings of modern geopolitics often underemphasized resistance/rebellion and what has been called the 'geopolitics from below' (see In focus 8.1).

While resisting the temptation to romanticize 'resistance' per se, critical geopolitical writers have sought to recover the complexities of global political life, including the role of individuals and societies who challenge hegemonic powers such as the United States. These histories and geographies of resistance/counterhegemony can be characterized as a form of anti-geopolitics (Routledge 1998, 2003a, 2003b) and dissident geographies more generally (Blunt and Willis 2000).

Following on from the writings of Antonio Gramsci (1971), anti-geopolitics is informed by a belief that the intellectual and cultural hegemony of an elite should not simply be accepted and even naturalized. Accordingly, 'anti-geopolitics can be conceived as an ethical, political and cultural force within civil society i.e. those institutions and organizations that are neither part of the processes of material production in the economy, nor part of state-funded or state-controlled organizations. . . . [Anti-geopolitics] challenge[s] the notion that the interests of the state's political class are identical to the community's interests' (Routledge 1998: 245). In other words, anti-geopolitics does not assume that the practical geopolitical reasoning of national elites reflects the interests and wishes of individuals and civil society. An important implication of this strand

In focus 8.1: Global protests against the 2003 Iraq war

On the weekend of 15–16 February 2003, 10 million people took to the streets to protest against the possibility (at that stage) of a US-led assault against Iraq. Co-ordinating events through the Internet, protestors not only held rallies in the public spaces of major cities such as London and New York but also used a variety of tactics such as jamming the White House switchboard and overwhelming the White House official Internet site. Seven hundred theatre groups were involved in simultaneous performances of the anti-war play, *Lysistrata*. Children as well as adults were involved in co-ordinated protests. British children joined a huge protest rally in London, Italian children blocked trains carrying US military personnel and Irish children gathered around Shannon airport to condemn the presence of American military aircraft using the facilities for refuelling purposes. Other European and American protestors travelled to Iraq as so-called 'human shields' in an effort to prevent further military strikes against Iraq (Anglo-US military action had been fairly continuous since the ending of the 1991 Persian Gulf conflict).

Protestors not only condemned the proposed military strikes as imperial in nature but also drew attention to the prevailing politics of oil. The US consumes 26 per cent of the world's oil supply and Iraq is widely believed to possess the second largest oil reserves in the world after Saudi Arabia. Moreover, President Bush is deeply implicated in the US oil industry as he was a former director of Harken Energy Corporation and Vice-President Cheney was CEO of the Halliburton energy company. During the 2000 presidential election, Bush and the Republican Party received millions of dollars in donations from oil companies. Protestors therefore poured scorn on claims that Iraq's oil reserves were not part of Anglo-American strategic thinking and their desire to overthrow the regime of Saddam Hussein.

of work is to move the centre of attention away from the territorial state to other forms of political entities such as social movements, anti-colonial groups and anti-globalization activists. It provides a timely reminder that political life is not exclusively defined by the nation-state.

Using the work of the British geographer Paul Routledge (for example 1996, 1998), this chapter explores the realm of anti-geopolitics, starting with an appraisal of anti-colonial movements and individuals and their methods of resistance against the political and geographical practices of European colonialism. Next, the Cold War is investigated to establish how dissident intellectuals and movements in Eastern Europe and the Third World strove to resist the superpowers and their attempts to impose their competing political visions on the world. Thereafter, recent expressions of resistance by anti-globalization movements are considered in order to demonstrate that they (and associated social movements) can and do operate across local, regional and global boundaries. Recent demonstrations against the World Trade Organization (WTO) and G8 summits in locations such as Seattle, Davos, London and Genoa illustrate how resistance is growing towards global institutions and large states and multinational corporations. The concluding part of the chapter acknowledges that implicit within, for example, the anti-globalization movement is a critique of the existing international territorial and economic order. Regardless of the location of resistance, Blunt and Willis have contended that, 'dissident geographies [like anti-geopolitics] . . . all share a political commitment to overturn prevailing relations of power and oppression' (2000: viii). The struggle remains, however, for critics to identify the ways in which hegemonic common sense (this might include the ideologies of trade liberalization and open markets) is constructed and to consider how alternative viewpoints can be mobilised.

Colonial anti-geopolitics

Definition Box: Colonialism and Imperialism

Colonialism refers to the systematic policy of acquiring and maintaining colonies for the purpose of political control and economic exploitation. Since the fifteenth century, European powers such as Britain, France, Belgium, Spain, Portugal, Germany and the Netherlands explored, conquered and appropriated most of the world, apart from places such as parts of China, Japan, Thailand and Saudi Arabia. **Imperialism** refers to a system of rule which enabled colonies to be brought under the control of the metropolitan power. Colonialism and imperialism were often justified on the basis that the European powers were helping to 'civilize' 'backward' peoples.

It is important to recognize two further dimensions of imperialism and colonialism. First, colonial and imperial values and practices may persist despite the formal ending of empires and colonies. Second, terms such as 'cultural imperialism' have been used to imply that countries such as the United States and the cultural practices and values (for instance, in the form of television programmes such as *Dallas*) originating in them can exercise a powerful grip on other countries, especially in the Global South. Moreover, in economic and political terms, large states such as the United States may enjoy 'spheres of influence' in geographically proximate regions such as Latin America.

The formal dissolution of European empires in the post-1945 period marked a fundamental change in global politics as a raft of newly independent states joined comparatively well-established states in Europe, North America and Latin America. The demise of European colonialism was unquestionably aided by the emergence of the two superpowers, the Soviet Union and the United States. As both unreservedly condemned colonialism, the end of the Second World War presented an opportunity to establish a new world order with due emphasis on the right of all peoples to self-determine their futures. Despite fiery outbursts against one another thereafter, the United States and the Soviet Union did share some political common ground. They had both fought in a global war against European fascism and their collective views were unquestionably shaped by revolutions against European colonialism (United States in 1776) and monarchical authoritarianism (Russia 1917).

The newly created United Nations stipulated in its 1945 charter that all forms of colonialism should be eradicated. The two major imperial powers, France and the United Kingdom, as part of their post-war economic transformation, had to agree to dissolve their imperial portfolios in return for financial assistance from the United States in the form of Marshall Aid. However, imperial break-up did not occur smoothly; imperial powers were at times reluctant to relinquish their claims over territories (for example, France over Algeria and Indo-China) or confronted with armed resistance by anti-colonial movements (for instance, against Britain, in Kenya, Aden, Malaysia and Cyprus; and see In focus 8.2).

Other imperial powers such as Portugal in Africa and the Netherlands in South East Asia were also to witness long and often bloody losses of imperial possessions.

Anti-colonial movements in particular called into question the apparently noble imaginative geographies of colonialism and imperialism. The vision of empire building as a selfless activity designed to 'modernize' or 'civilize' native populations was roundly rejected. Colonial representations of cultures and peoples as 'backward' and/or 'primitive' were actively contested – an attitude perhaps best immortalized by the Indian Independence leader Mahatma Gandhi when he said that 'British civilization' was a good if unfulfilled idea! Gandhi, who had spent part of his life fighting against racial discrimination in South Africa before returning to India, was well placed to offer an

---

In focus 8.2: Anti-colonial resistance against imperial Britain

In the post-1945 period, successive British governments used terms such as 'terrorists' to describe those engaged in subverting colonial rule. The label 'terrorist' became useful in the justification of the violent suppression of such acts of resistance by British armed forces. The most notable examples include:

1. Zionist insurgency in Palestine 1944–7, which witnessed Jewish and Zionist attacks against British military and civilian establishments prior to Israeli independence in 1948.
2. The Malayan Emergency 1948–60, a long-term campaign against the activities of the Malayan Communist Party. Malaysia eventually won independence in 1960.
3. The Mau Mau Uprising in Kenya between 1952 and 1960, involving the struggles of Kenyans to gain greater land reform and autonomy vis-à-vis the colonial administration. Kenya gained independence in 1962.
4. EOKA's (National Organization of Cypriot Fighters) campaign against the British in Cyprus in 1955–9 for the purpose of promoting a political union between Cyprus and Greece. Cyprus gained independence in 1960 but a Turkish invasion in 1974 later led to the division of the island between the Greek- and Turkish-speaking communities.

Sources: Furedi 1994 and Carruthers 1995

assessment of the imperial authority of the British. His followers pioneered strategies such as extensive civil disobedience, which were later to be adopted with considerable success by charismatic leaders such as Martin Luther King and by the US Civil Rights movements in the 1960s. Tragically, Gandhi was blamed by some for the partition of India in 1947 following independence from imperial Britain and was assassinated by a Hindu extremist in 1948.

Colonialism therefore depended on a worldview of a civilized Europe co-existing uneasily with the uncivilized spaces of Africa, Asia and the Middle East. The novelist Joseph Conrad's infamous depiction of the Congo (which until its independence in 1960 was under the imperial control of Belgium) as the 'heart of darkness' was one such representation of colonial spaces. Even after independence, Central Africa was still considered to be beyond the realm of reason and civilization, and shockingly, during the 1994 Rwandan genocide some Western media commentators resurrected the phrase 'the heart of darkness' to imply that the horrendous level of violence was due in part to the 'backwardness' of the local population. Genocide was naturalized.

This injudicious labelling of place stands in awkward contrast to the fact that the worst genocides in twentieth-century history occurred in Central and Eastern Europe (the Holocaust) and South Eastern Europe (the Armenian genocide of 1915–16). The Soviet Union under Stalin also perpetrated countless atrocities against Jews, political opponents and dissidents, and mass murder was a defining feature in the destruction of the former Yugoslavia in the 1990s. Genocide, like the concentration camp (first used by the British in South Africa in 1899), was arguably invented in Europe. Former imperial powers (Belgium and France) did precious little to prevent the outbreak of the Rwandan genocide. The absence of an effective United Nations during the 1994 genocide was also lamentable, as over 1 million perished and millions more have been killed and/or injured in continued fighting within the Democratic Republic of Congo (see Chapter 7).

Challenging colonialism was thus not only a struggle against the matériel power of imperial Britain or France but also a defying of the cultural power of the colonial imagination. Why was a place such as the Congo still described as a 'heart of darkness'? Post-colonial scholars such as the late Palestinian academic and activist Edward Said and the Martinique-born doctor and anti-colonial writer Frantz Fanon have highlighted the insidious power of colonialism. They have argued that it is far easier to dismantle the formal apparatus of colonialism (such as border posts, administrative buildings and imperial troops) than it is to change the thinking habits of a lifetime. Power, knowledge and representation lie at the heart of the imperial condition, as does the struggle over imaginative geographies.

Frantz Fanon and anti-colonialism

The Martinique-born doctor and psychiatrist Frantz Fanon was one of the intellectuals involved in the nationalist organization Front de Libération National (in English NLF), which was primarily engaged in anti-colonial resistance against the French in

Algeria. After a humiliating defeat in Vietnam in 1954, the reluctant French departure from Algeria was arguably one of the most brutal episodes in post-war European colonialism. In his famous book, *The Wretched of the Earth* (1961), Fanon articulates a view of colonialism as an inherently violent phenomenon in both a physical and an intellectual sense (see In focus 8.3).

In focus 8.3: Imperialism v. progressive socialism

The writings of Fanon and other anti-colonialist writers such as Che Guevara drew attention to a world divided between Euro-American imperialism and progressive socialist governments such as that of Cuba. Progressive socialism refers to a body of thought and practice concerned with the promotion of public ownership of the means of production and a commitment to social justice for everybody regardless of gender, age and ethnicity.

In the early 1960s, Third World hostility to the United States was mounting not only because the Kennedy administration was involved in the assassination of the Congolese president and progressive socialist Patrice Lumumba in 1960, but also because Afro-Americans were being denied their basic human rights. The United States was seen by some Third World anti-colonial writers as hypocritical, since neo-colonial relations persisted well after decolonization. Given the violence of the Cold War, formal independence looked a hollow promise, and socialists advocated world revolution.

As Chapter 3 demonstrated, the creation of the Non-Aligned Movement in the 1960s owed a great deal to the frustration felt by post-colonial states at their lack of political and economic development in the midst of a highly costly and divisive Cold War.

The brutality of both sides during the wars of independence in Algeria (1954–62) produced countless massacres and acts of torture. During this period, Fanon provided medical and psychological support to the NLF. Although independence was gained in 1962, the cost of the struggle was substantial: over 12,000 French soldiers were killed and there were perhaps half a million Algerian military and civilian casualties. Algeria's post-colonial condition has worsened further since the 1991 presidential elections, which witnessed the emergence of the Islamic Salvation Front (FIS). The FIS promised to create an 'Islamic' state but this proved unacceptable to the Francophone/secular armed forces, who intervened and imposed their own president, Chadli Benjedid. Forty years after independence, Algeria remains immersed in a violent civil war, which has claimed over 150,000 lives as rival factions fight over the future shape of Algerian politics and culture. France and more recently the United States continues to intervene by supporting the Algerian armed forces in their so-called 'war against Islamic terror'.

Intellectually, Fanon contended that colonialism was predicated on the assumption that the colonized were always considered inferior to the colonizer (see Young 2003). The colonized are dehumanized within colonial discourse, and their right to represent

themselves tends to be circumscribed by the colonial power. Thus in order to decolonize, the colonized must not only expel the physical presence of the oppressors but also
regain an ability to represent their own culture. Anti-colonial violence was legitimate
because colonialism was a violent process. In *A Dying Colonialism*, Fanon reconsidered
the Algerian revolution and concluded:

> What we Algerians want is to discover the man behind the colonizer; this man who is
> both the organizer and the victim of a system that has choked him and reduced him to
> silence. As for us, we have long since rehabilitated the Algerian colonized man. We have
> wrenched the Algerian man from a centuries-old and implacable oppression. We have
> risen to our feet and we are now moving forward. Who can settle us back in servitude?
> We want an Algeria open to all, in which every kind of genius can grow (Fanon 1967
> cited in Blunt and Willis 2000: 183).

Fanon's appeal, five years after actual independence, sits uneasily with the contemporary situation of Algeria, enveloped in a bitter civil war where women and children
in particular are paying a high price as Algerian men struggle to define the civic and
religious nature of the state.

Edward Said and the imaginative power of Orientalism

As a Professor of English and Comparative Literature at Columbia University in New
York, the late Edward Said has been at the forefront of academic and political studies
of the so-called Palestine problem (Fig. 8.1). Born in Jerusalem and educated in colonial Egypt and the United States, Said had experienced and been shaped by imperial
encounters. As with much of the contemporary Middle East, the imperial imprint of
the British, French and Americans looms large in the social and political geographies
of Egypt, Israel and Palestine. Imperial powers not only drew lines in the sand (not to
mention rock and ice) but also shaped lines in the mind.

Through a series of publications stretching from *Orientalism* (1978) to his memoir
Out of Place (1999), Said demonstrated how a series of 'imaginative geographies' have
structured Western representations and understandings of the Middle East (Orient)
and the 'West'. British, French and American imaginings and representations of the
Orient are claimed to have been of great cultural and intellectual significance (see
In focus 8.4).

The 'imaginative geographies' of the Orient and the Occident are intimately linked
to one another. The West was once considered the norm which other cultures and
peoples should emulate. Subsequently, this view was exposed as inherently harmful to
the people not only of Palestine but of the entire Middle Eastern region precisely because
of their alleged asymmetric relationship to Western cultures.

Said contends that these understandings of the Orient as backward, mysterious
and uncivilized continue in the allegedly post-colonial era and have contributed to the
legitimating of the Israeli occupation of the West Bank/Palestine. Democratic, settled
and civilized Israel is thus, according to Said and many other Arabic writers, compared

Figure 8.1   The late Edward Said. An academic and an activist, he was the author of numerous influential writings, including *Orientalism* (1978)
Photo: PA Photos

In focus 8.4: Edward Said and the invention of the Orient

The Orient was almost a European invention, and had been since antiquity a place of romance, exotic beings, haunting memories and landscapes . . . The Orient is not only adjacent to Europe; it is also the place of Europe's greatest and richest and oldest colonies, the source of its civilizations and languages, its cultural constant, and one of its deepest and most recurring images of the Other. In addition, the Orient has helped to define Europe (or the West) as its contrasting image, idea, personality and experience. Yet none of this Orient is entirely imaginative. The Orient is an integral part of European material civilization and culture (Said 1978: 1–2).

In focus 8.5: Media representations of the Israeli-Palestinian struggle

The general picture is that Israel is so surrounded by rock-throwing barbarians that even the missiles, tanks and helicopter gunships [supplied by the United States] used to 'defend' Israelis from them are warding off what is essentially an invasive force. . . . In the US media, Zionization is so thorough that not a single map has been published or shown on television that would risk revealing to Americans the network of Israeli garrisons, settlements, routes and barricades, which crisscross Gaza and the West Bank. Blotted out completely is the system of Areas A, B, and C, which perpetuates military occupation of 40 per cent of Gaza and 60 per cent of the West Bank . . . The censorship of geography, in this most geographical of conflicts, creates an imaginative void . . . in which all images of the conflict are decontextualised (Said 2000: 45–6).

See also Chapter 4 of this volume, on popular geopolitics.

---

favourably to backward and unstable Palestine. Accusing American supporters of Zionism of exercising a virtual veto on discussions of Israel and Palestine within the United States, Said contends that these 'Oriental' representations of Palestine permeate popular culture, especially television reporting (see In focus 8.5).

Since 1948, when Israel was created in part through anti-colonial terrorism directed against the British and partly as a consequence of a terrible genocide against the Jewish people, it has benefited from continual military, financial and political support from the United States. Its nuclear weapons programme (dating from the 1960s) was actively encouraged by the United States and France, and unlike the 'axis of evil' states such as Iraq and Iran, Israel has never been condemned for holding substantial stocks of weapons of mass destruction (see Chapter 9). In contrast, the Palestinians have received no support from the United States but have been condemned for political terrorism and cultural backwardness. Israel continues to occupy Palestinian territory, and thus the role of the dissident, for people like Edward Said and others, is to expose and contest the enduring power of colonial discourses and their specific impact on colonized peoples such as the Palestinians.

Decolonization and the Cold War: the disappointment of independence?

The experience of decolonization became a global phenomenon by the 1960s, as large swaths of Africa and Asia completed a process that was arguably initiated by the United States in the eighteenth and Latin America in the nineteenth centuries. When in 1960 the British prime minister Harold Macmillan announced to the South African Parliament that the 'winds of change' were blowing through the African continent he underestimated the extraordinary complexity of decolonization. Decolonization did not occur in a geopolitical vacuum, as the Soviets and the Americans struggled to establish their authority over a so-called Third World of mainly post-colonial states. Anti-colonial struggles were sometimes actively supported by the superpowers, who themselves became embroiled in colonial struggles. The American involvement in Vietnam in the 1960s and their eventual defeat followed an earlier defeat of the French armed forces in Dien Bien Phu in 1954. In other words, French colonialism was replaced by an ill-fated attempt by the United States to impose its military will on South East Asia. After the American withdrawal between 1973 and 1975, a divided Vietnam was eventually unified into a new post-colonial state.

Decolonization, however, produced deeply unsatisfactory results for many newly independent states. In Africa and South Asia, for example, post-colonial states inherited colonial boundaries that bore no relation to ethnic, religious and cultural borders. The 1947 partition of India and Pakistan had tragic consequences as Muslims and Hindus swarmed across the border after finding that they were on 'the wrong side'. Muslims found that they were no longer welcome in Hindu-dominated India, and Hindus were considered unwelcome in Pakistan. Over 500,000 died as Sikhs and Hindus clashed with Muslims in the Northern provinces of Punjab and Kashmir. As one writer, Patrick French, has graphically concluded:

> The nature of the political settlement in 1947 had a calamitous impact on the subcontinent, leading to the reciprocal genocide and displacement of millions of Hindus, Muslims and Sikhs, three Indo-Pakistan wars, the blood-drenched creation of Bangladesh, and the long-term limitation of the region's global influence. Although more than a fifth of the world's population presently lives in the territory of Britain's former Indian Empire, continued internal conflict has left South Asia with little cohesiveness and minimal international clout. Nearly half of all Pakistani government expenditure still goes on the cold war with India, focussed on the running sore of Kashmir (French 1998 cited in Blunt and Willis 2000: 177).

Elsewhere, new governments in Africa such as Nigeria (independent in 1960) were suddenly expected to administer a national territory which bore no relation to existing tribal boundaries in West Africa. For some minority ethnic groupings, independence did not entail an improved position within the post-colonial state, because a new generation of political elites became adept at seeking advantage from existing tribal and ethnic boundaries and groupings.

Thus it has been contended that the ceding of formal colonialism did not necessarily entail the ending of well-entrenched colonial practices and attitudes. Some critics maintain that new forms of colonialism (**neo-colonialism**) came into existence and that the superpowers used the Cold War as an excuse to impose or support 'friendly' Third World leaders and governments who exploited their own countries. One of the most infamous examples was the rise of President Mobutu of Zaire (until 1970 the country was known as the Republic of Congo), who with the help of a CIA-backed coup took over governance in 1965. Given Zaire's position within central and southern Africa in conjunction with its ample supplies of diamonds, cobalt and copper, the United States was prepared to overlook outbreaks of ethnic violence and Mobutu's massive theft of government monies so long as he remained a reliable strategic ally in the region. For over 30 years, the people of Zaire were forced to endure this despot, despite the formal rhetoric of the United States urging all Third World governments to embrace the ideals of liberty, democracy and the free market. Zaire had previously been a Belgian colony and subject to a costly civil war in the early 1960s as the copper-rich province of Katanga attempted to leave the then Republic of Congo. Another example of alleged neo-colonialism is Indonesia, where the United States supported the coup by General Suharto in spite of extensive massacres of the country's inhabitants culminating in the illegal invasion of the former Portuguese colony of East Timor in 1975. Indonesia was previously a Dutch colony. In both cases, the struggles for decolonization produced violent post-colonial states, governed by autocratic leaders who were openly supported by the United States as part of their strategy to extend their influence across the Third World during the Cold War. Massacres and genocide were tolerated in a political culture made 'mad' by the extreme dangers posed by rival ideologies and WMD.

The struggle for independence was only one part of the process of decolonization. Independence, as we have seen, did not guarantee political and social stability or vouchsafe a flourishing democracy. India was one of the major exceptions to the general rule that post-colonial states could not generate a working democratic culture. Despite their promises on the eve of independence, many post-colonial leaders were unable to produce either a capitalist or a socialist paradise. President Castro's socialist Cuba, however, managed to survive (since 1959) despite extraordinary levels of hostility from the neighbouring United States. New states found themselves embroiled in a Cold War not of their making, and participants in an international political and trading order which, as many Third World scholars contended, actively prevented them from making substantial economic improvements. Active backing from the superpowers proved a mixed blessing for some states. Israel was able to develop both politically (a democratic parliament) and economically (a market economy) while others had to endure vainglorious and violent dictatorships and governments that simply plundered the natural resources of their countries. Nominally socialist Third World states were not necessarily any better than their capitalist counterparts, as the Pol Pot regime in Cambodia demonstrated when it murdered over 1 million of its citizens in the 1970s.

Post-colonial transformation did not necessarily mean that colonial discourses would dissipate. It has proved hard to dismantle the kinds of intellectual and imperial structures, which shaped relations between the colonizer and the colonized. Post-colonial states in Africa, the Middle East and Asia continue to be depicted as 'underdeveloped', 'immature' and/or 'backward'. During the Vietnam War, the Vietnamese were frequently depicted as passive, lazy and vulnerable to outside control. A more recent case in point is the persistent American political and media representations of Iraq and its inhabitants during the 2003 war against Saddam Hussein's regime. As the British novelist Jonathan Raban remarked, 'On television, the Iraqis themselves have been relentlessly feminised and infantilized, exactly along the lines described in Said's *Orientalism*. They are the Little People: all heart and no head, creatures of impulse and whim, not yet grown up enough to make rational decisions on their own behalf' (Raban 2003: 6).

The George W. Bush administration subsequently appointed civilian and military administrators to manage Iraq and, as with post-1945 Japan and Germany, they will be responsible for writing a new constitution for the Republic.

Cold War dissent and anti-geopolitics

Decolonization in the Third World and Soviet-inspired repression within Eastern Europe were two of the most striking features of the Cold War period. Both superpowers were active in promoting their geopolitical and geo-economic interests in regions which they perceived to be their 'backyards'. As Chapter 3 illustrated, Latin America was the region of greatest interest for the United States. For the Soviet Union, however, Eastern Europe was the 'backyard' and the Soviet leadership extended their control from the Baltic to the Black Sea. Military and economic alliances such as the Warsaw Pact and COMECON were one element in the web of influence. Socially and culturally, Eastern European societies were under the tight control of local communist parties and uniformity was secured through oppressive communist party leaderships and the feared secret police forces. Ideological consent was promoted and facilitated by communist dogma relating to public ownership and unquestioning loyalty to the communist party and the ideals of socialism.

For much of the Cold War period, everyday life in Eastern Europe was tightly regulated and citizens had little opportunity to express dissent against the ruling communist regimes. Why was this the case? Three factors might be cited as part of an explanation. First, the communist authorities were deeply fearful of allowing any capitalist influences to 'contaminate' their societies for fear that the people would be corrupted by Western decadence and material greed. Second, by restricting flows of information and commodities, the communist parties sought to prevent any critical debate about how and with what success socialist goals were being achieved. Economic planning, for example, was more often than not characterized by fictitious statistics because it was considered imperative that socialism in Eastern Europe was shown to be working successfully. Third, some have contended that individual leaders such as Stalin were able to draw on a long tradition of authoritarian rule from Ivan the Terrible onwards to justify their style of governance.

How was repression secured? It used to be said that in return for a guarantee of full employment and welfare ('from the cradle to the grave'), citizens had to forfeit any right to freedom of speech, association and movement (see Chapter 7). Control over public and even private expression was secured by coercion on the one hand and propaganda and public education on the other. Given the extensive network of secret police contacts within Eastern European families, there was always the fear that a member of one's own family would denounce one to the secret police for pro-imperialist and/or counter-revolutionary thoughts. Wives reported on husbands and sons betrayed their fathers, often on the basis of overheard 'private' conversations in the family home.

The British novelist George Orwell's depiction in *Nineteen Eighty-four* of a world regulated by ministries of love and truth was not that far removed from the grim

In focus 8.6: Sport and the popular geopolitics of the Cold War

For much of the Cold War, sporting events such as the Olympic Games were highly charged affairs, as both superpowers viewed victory as an affirmation of their particular social and political systems. High-profile team events such as basketball and ice hockey were frequently laced with political tension as the Soviets and the Americans battled for the gold medal. Likewise individual sports such as boxing were often interpreted as contests between the two states rather than matches between two sports competitors. In 1980 the United States boycotted the Moscow Olympic Games in protest against the 1979 Soviet invasion of Afghanistan, and in 1984 the Soviets boycotted the Los Angles Olympic Games. Both sides accused one another of using performance-enhancing drugs, and in the aftermath of the Cold War it emerged that communist East Germany in particular was guilty of promoting drugtaking in an attempt (often successful) to accumulate gold medals in swimming, athletics and weight lifting.

On a more positive note, sport was also used to promote better diplomatic relations. Most famously, in 1972 the Nixon administration sent a team of table-tennis players to China in order to improve Sino–American relations. The term 'ping pong diplomacy' was invented and added to the diplomatic lexicon.

These kinds of events (and associated depictions of national prowess and ideological superiority) remind us that popular geopolitics also extends to the world of sport.

realities of everyday life for the citizens of Eastern Europe. The freedom to travel outside the socialist bloc (with the partial exception of Yugoslavia) was also tightly controlled, so that for citizens other than communist party officials, the most obvious opportunities for travel were afforded through sport and cultural pursuits such as ballet, chess and music. Like Nazi Germany in the 1930s, East European regimes invested heavily in sporting programmes, and public successes in high-profile events such as the Olympic Games were taken to be a vindication of communist governments and state socialism. Famous sports performers sometimes defected to the West. Tennis player Martina Navratilova secured US citizenship after she successfully claimed political asylum from her native Czechoslovakia, as did the Russian ballet dancer Rudolf Nureyev in 1961 after a visit to Paris (see In focus 8.6).

Resisting Cold War communism

Popular uprisings against Soviet-backed regimes in Eastern Europe did occur during the Cold War period. Most infamously, the Soviets used overwhelming military force in Hungary in 1956 and Czechoslovakia in 1968 after attempts to overthrow the communist regimes. Trade-union movements such as Solidarity and the Roman Catholic church were also actively targeted by the regime in Poland in the early 1980s. Dissident intellectuals and writers such as Alexander Solzhenitsyn were hounded by the secret police and some were sent to labour camps (gulags) in the Far Eastern Russian region of Siberia. Ironically, two of the most charismatic leaders of resistance, Vaclav Havel in Czechoslovakia and Lech Walesa of Poland were later to become their country's first presidents in the post-communist revolutions in the early 1990s. Walesa, with the support of the Catholic church and the new Polish pope, John Paul II, began to organize resistance against the regime, which was forced to apply martial law in an attempt to suppress Solidarity. However, the ailing Polish economy failed to meet the social and economic demands of millions of Poles, and Walesa and his supporters swept to power in 1990 following a popular uprising.

As someone who was imprisoned for nearly ten years by the Czech regime for his dissent, Havel understood well the power of the Eastern European communist regimes to shape public and private thought. The manufacture of consent was an essential element in the pseudo-legitimacy of Eastern European regimes. Citizens were systematically indoctrinated from a very early age and breaking out of this hegemonic grip required a complete dissolution of the intellectual and artistic strictures of Soviet-style totalitarianism. Havel and other dissident intellectuals such as Andrei Sakharov in Russia encouraged the development of a 'civil society' independent from the public culture sponsored by the regimes. 'Underground' culture in the form of newspapers, music groups and secret meeting groups began to flourish in Czechoslovakia, especially after widespread public revulsion at the violent suppression of the so-called 'Prague Spring' uprising in 1968. Similar outpourings of dissent and revolt in Western European cities such as London, Paris and Rome witnessed street protests against the inequities of capitalism, rigid social hierarchy, institutionalized racism and bland public culture.

One of the most significant movements to emerge in the post-1968 period was the Czech-based Charter 77, with Havel as a founding member. The group not only focused attention on the (lack of) protection of human rights in Eastern Europe but also began to articulate alternative views on social and political organization. Arguably, Charter 77 laid the foundation for increased demands for genuine participation in public life, especially following Soviet leader Mikhail Gorbachev's policy of glasnost (openness) and perestroika (restructuring) in the mid-1980s. In the last years of the Cold War, citizens across Eastern Europe took to the streets demanding greater political, cultural and social freedoms, with dissident groups playing a key role in facilitating this change.

These dissident movements in Eastern Europe forged significant links with peace and social movements in Western Europe. Anti-nuclear movements such as CND and European Nuclear Disarmament (END) protested throughout the 1970s and 1980s against the deployment of nuclear weapons in Western and Eastern Europe by NATO and Warsaw Pact forces. CND and END attempted, with a large measure of success, to create a network to facilitate trans-European solidarity and identity. Intellectuals such as the British philosopher Bertrand Russell and the historian E. P. Thompson became high-profile campaigners and articulated a dissident geography of the Cold War. Condemning the 'frozen' geographies of East and West, Thompson appealed for an ending of European divisions. The massive build-up of conventional and nuclear weapons not only generated insecurity but also served to legitimate the interests of the superpowers by confirming that the other side was an unrelenting threat.

With the election of right-wing governments in the United States (Reagan in 1980) and the UK (Thatcher in 1979), the peace movements recognized that the underlying geopolitics of the Cold War needed to be actively contested lest a Third World War erupt within Europe. Thompson and others urged the peace movements of Western Europe and the dissident groupings of Eastern Europe to promote international solidarity and peace by revitalizing civic culture and expressions of protest. One of the most disturbing features of the latter stages of the Cold War (i.e. the early to mid 1980s) was the manner in which many administrations spoke ominously about the dangers posed by the other side and the need to maintain large stocks of nuclear missiles in order to secure collective security. The principles of CND were not widely accepted within Western European electorates. The British Labour Party, which adopted the principle of unilateral disarmament, lost two successive elections in 1983 and 1987.

The subversion of Eastern European regimes culminating in the demolition of the Berlin Wall in November 1989 (see Fig. 8.2) effectively ended the Cold War geopolitical divisions within Europe as citizens in the East eagerly embraced the lifestyles and political practices of the West. On a visit to Estonia in June 2001, for example, it struck me that Estonian teenagers looked indistinguishable from their 'Western' European contemporaries. The dissolution of the Soviet Union in 1991 confirmed that the Cold War was over, and both sides now (2003 onwards) collaborate closely in the 'war against terror'. While rejecting some of the grander claims that this moment spelled the 'end of history', there can be little doubt that ideas and practices associated with democracy and economic neo-liberalism have taken hold in many areas of the world including the former Eastern European bloc and many parts of the so-called Third World (see Chapter 1).

Globalization of dissent

The ending of the Cold War assured in the short to medium term (though not necessarily for ever) that the type of communism that existed under the Soviet Union and its allies will be unlikely to find widespread favour again. The human and environmental cost of state socialism in Russia, China and Eastern Europe was immense. With the exception of North Korea and Yemen, most former and/or communist countries are moving towards an economic doctrine based on economic liberalism. Fundamentally, it is assumed that the interests of states and the international trading system are best served by having open markets, but restrictions remain, including national subsidies, welfare programmes and price controls. In order to be efficient, markets need to be free from state interference and national compromises such as wage or pricing agreements. With the backing of the United States, the international financial and trading system headed by institutions such as the International Monetary Fund (IMF) and the World Trade Organization (WTO) seek to remove all barriers to free trade and capital. It is further contended by neo-liberal economists that transnational corporations

Figure 8.2 The demolition of the Berlin Wall in November 1989 heralded the dramatic end of the Cold War in Europe
Photo: PA Photos

also need to be freed from 'excessive' restrictions regarding the movement of labour and capital. As part of this political-economic creed, national economies need to be deregulated and state regulation reduced in favour of private enterprise and decreased public spending.

While emphasis is inevitably placed on the free movement of capital and commodities, states have been urged to control their public spending in order to avoid inflationary pressures developing within national and international economies. The mantra of economic liberalism is not always respected in practice, however. Western governments such as America and the European Union condemn Third World states for protectionism while they insist on protecting their own agricultural and industrial sectors. At the same time, while Euro–American governments attempt to protect their national economies, they strive to exercise tighter control over national territory by imposing strict regulations on the movement of people across borders. It is remarkable, as we have noted earlier, how money is allowed to flow across the world, often with the minimum of fuss, while people searching for new economic opportunities are automatically condemned as 'illegal' migrants. High-level spending by the US Border Patrol is just one example of a state's determination to control the flow of people from another state (in this case Mexico) even if most officials concede that hermetically sealing a 2,000-mile border is well-nigh impossible.

Regional economic agreements such as the North American Free Trade Agreement (NAFTA) between the United States, Canada and Mexico illustrate well how states, markets and people are complicit within a highly unequal process of production and exchange. According to the disciples of neo-liberalism, NAFTA should promote free trade and mutual competitive advantage. The three states would be free to concentrate on the production of goods and services in which they enjoy a comparative advantage once all forms of restrictions are removed. In reality, the removal of obstacles to free trade has meant in particular that large American corporations have been able to exploit cheaper labour and laxer environmental restrictions in Mexico while seeking to prevent Mexican labour from crossing the border in search of opportunities within the United States. Many Mexicans have been impoverished, and the poor, peasants, women and indigenous peoples have most keenly felt the impact of transnational liberalism. At the same time, the Mexican government, with its long history of default on loan repayments, has been actively encouraged to cut expenditure with serious repercussions for public health, welfare and unemployment. Ironically, many American producers remain dependent on cheap (and often illegal) Mexican labour for harvesting crops such as strawberries by hand (see Chapter 3). In belated recognition of their role in sustaining the US borderland economies, President George W. Bush has proposed an amnesty on illegal Mexican labourers living in the United States. Critics contend, however, that this proposal is designed to secure a greater share of the so-called Hispanic vote in the forthcoming 2004 presidential election.

Resisting globalization

It is important to note that these developments associated with globalization and regionalism have not been accepted uncritically (see In focus 8.7).

Within North America, the Zapatistas in the Chiapas region of southern Mexico (close to the border with Guatemala) have provided a powerful example of resistance to economic globalization. Named the Zapatista National Liberation Army (ZNLA) after the revolutionary leader Emiliano Zapata (1879–1919), the Zapatistas were inspired by a Mexican struggle to overthrow entrenched privilege and wealth in the 1910s and 1920s. Their strategies and tactics can be thought of as a kind of anti-geopolitics precisely because they challenge the dominant vision of corporate United States/ Mexico/ Canada and the wider international financial and trading institutions (see O Tuathail 1997). Composed of predominantly indigenous peoples (Mayan), the Zapatistas demanded not only the ending of NAFTA but also the democratization of Mexican civil society. They launched their guerrilla movement on the day that NAFTA formally came into effect and occupied the main city of the Chiapas region. Power supplies were disrupted and car bombs were set off in Mexico City and the tourist resort of Acapulco. As one rebel noted: 'We have nothing, absolutely nothing . . . not decent shelter, nor land, nor work, nor health, nor food, nor education. We do not have the right to choose freely and democratically our officials. We have neither peace nor justice for ourselves and our children. But today we say enough!' (cited in Rogers 2002: 78).

Almost immediately it became apparent that the movement was eager to globalize its struggle (Routledge 1998, Rogers 2002). Under the leadership of Subcommandante

In focus 8.7: Two types of dissenters to globalization

The anti-global groups that exist around the world are diverse and espouse different political, economic and cultural priorities. However, it is possible to distinguish between two types of critics.

1. **The national/regional protectionists.** This group blames globalization for their domestic countries' or regions' woes such as unemployment, the erosion of social patterns and the power of global investors. They condemn the loss of national self-determination and the destruction of national and/or regional cultures and their concern is primarily with protecting their own citizens rather than promoting a more equitable global economic and political order. Membership of this group might include right-wing political figures such as Pat Buchanan in the United States, nationalist parties such as France's National Front and Austria's Freedom Party, and Osama bin Laden with his vision of a 'pure' Islam (and a corresponding Islamic community, the Umma) untouched by the influence of the United States and Northern-led globalization.

2. **The universal protectionists.** This group is also highly critical of the divisive nature of globalization but seeks to promote a more equitable relationship between the North and the Global South. Membership of this group is diverse and includes progressive political parties, NGOs, private citizens, trade unions and transnational networks. In general terms, they seek a new global order based on universal human-rights protection, greater financial equity, fair trade and the recognition of labour and women's issues. Membership of this group includes the Zapatistas (Mexico), the People's Global Action and numerous representatives from trade unions, farming groups and charities such as the Korean Advanced Agricultural Federation. In the case of the latter, one of its members, Lee Kyung-hae, killed himself at the September 2003 meeting of the WTO in protest at the gross inequalities within global trade.

Source: Adapted from Steger, 2003: 114–15

Marcos, the international media and the Internet was deployed in order to publicize the Zapatistas' demands to a wider audience. Their actions caused international panic throughout the financial markets, which responded badly to the uprising. The Mexican government was urged by the United States and the international financial community to restore a semblance of order. For the Zapatistas, the purpose of the revolt was to expose not only the inequities of global capitalism but also the manner in which the national sovereignty of Mexico had been 'sold' to the needs of international investors and transnational corporations. Marcos, through a series of Internet-based communiqués, highlighted how the Chiapas region, and in particular the Mayan population, was being subjugated to the demands of national and international economic interests. In so doing, he drew attention to how a poor region of Mexico was being colonized not only within Mexico but also more widely by external forces that were unaccountable. The deployment of the Mexican military forces within the Chiapas region merely illuminated still further the impoverishment of local democratic and economic rights. The military stood accused of simply responding to pressures imposed by the international financial community rather than the national government in Mexico City.

Resistance to global capitalism and economic neo-liberalism is not restricted to the Chiapas region of Mexico. A transnational collection of social movements has mobilized against the apparently unrestricted movement of capital and commodities. Far from being poorly educated and/or ignorant, protestors are well informed and have demonstrated skill in exposing the inequalities of global capitalism (see In focus 8.8).

International organizations and national governments stand accused of promoting a form of global capitalism which marginalizes the poor, women, indigenous groups and industrial workers. Cult figures such as French farmer José Bové led the opposition to high-profile corporations such as the food retailer McDonald's. Local acts of resistance such as Bové's high-profile destruction of a McDonald's restaurant under construction in Larzac in South West France are increasingly linked to global expressions of dissent. Until his well-publicized act of destruction in August 1999, Bové was a little-known sheep farmer from Millau and one-time leader of a French agricultural trade union, Confédération Paysanne. Despite expressions of sympathy across France and the wider world, he was jailed for property damage and accused of destroying a cache of genetically modified corn in local fields. Interestingly, Bové's anger, while targeted against a US chain of fast-food restaurants, was inspired by his hostility against a US tariff imposed on Roquefort cheese following an earlier refusal by the EU to import American hormone-treated beef. Presumably his local McDonald's restaurant would have used such beef.

In focus 8.8: Movimento Sem Terra (MST, Movement of the Landless Workers)

Created in Brazil, the MST has campaigned for radical land reform since the 1980s. Armed with the slogan 'Occupy, resist and produce', it has sought to mobilize 12 million landless workers and urged the occupation of land, especially where it is unoccupied and uncultivated. Alarmed by MST's success, the Brazilian government was given $150 million in special aid by the World Bank in order to develop a market-based land-reform scheme allowing landless people to borrow money for the purchase of property. This scheme was heavily supported by landowners who were worried that the regulatory power of the state was being undermined by a grassroots organization which bypassed the Brazilian federal and regional governments. Following the election of the trade-union leader Luiz Inacio da Silva (known universally as Lula) in November 2002, many landless peoples and indigenous Indian groups are hopeful that land reform might be urgently addressed; Brazil has the most unequal land distribution in the world. Three per cent of Brazil's population owns 70 per cent of the total available land and many indigenous Indian groups have been violently removed from their homes in the Amazonian basin by mining, ranching and timber corporations. The MST has close links with the People's Global Action group which campaigns for global justice, especially in matters such as land ownership.

Anti-globalization protest

The globalization of resistance has inspired a new generation of people to take to the streets across the world to protest against the corporate agenda of contemporary global capitalism (see Fig. 8.3).

The most high-profile acts of resistance have coincided with meetings of the WTO and the so-called G7 (the seven largest economies of the world but periodically augmented by Russia to create the G8). The WTO in particular stands accused of imposing an unfair rules-based international trading system, which is insufficiently attentive to global justice and fairness. Groups such as People's Global Action (created in 1998) illustrate how resistance to economic globalization is worldwide, as support has come from India, Mexico and Brazil as well as from European and North American groups. But there has been a seismic shift in the geographical distribution of protest. Previously, protests against institutional organizations such as the IMF were largely confined to the Third World, as people there took to the streets to protest against Structural Adjustment Programmes in the 1980s. Notwithstanding acts of solidarity between protestors in the North and South, it was not until the 1990s that resistance against 'globalization' became more universal. As Joseph Stiglitz, the former chief economist at the World Bank, recognized in his book, *Globalization and its Discontents*:

> Virtually every major meeting of the International Monetary Fund, the World Bank and the World Trade Organization is now the scene of conflict and turmoil. The death of a protestor in Genoa in 2001 was just the beginning of what may be many more casualties in the war against globalisation. Riots and protests against the policies and actions by

---

Cologne, June 1999. Protestors gathered outside a meeting of the G8 Summit in Germany.

Seattle, November/December 1999. 50,000 people protested outside the third meeting of the WTO in the US city of Seattle.

Prague, September 2000. 10,000 protestors attempted to sabotage the annual meeting of the IMF and World Bank.

London, May 2001. Thousands of activists marched through the centre of London in protest against the WTO, the IMF and the unequal nature of globalization more generally.

Genoa, July 2001. 100,000 gathered in order to disrupt the annual summit of the G8 in Italy. One protestor was shot dead by Italian police and violent battles unfolded on the streets between activists and police forces.

Cancun, September 2003. Thousands gathered to protest outside a WTO meeting in the Mexican resort town of Cancun. One South Korean farmer killed himself as crowds denounced the unequal nature of world trade.

Other meetings such as the 2001 WTO meeting in Doha were not subject to protest because such activities are banned by the state of Qatar. The 2002 G8 summit meeting was also spared high-profile protests because it was located in an inaccessible part of the Canadian Rockies.

Figure 8.3   Recent anti-globalization protests

In focus 8.9: Jubilee 2000 Campaign

The Jubilee 2000 Campaign was launched in London on 13 October 1997 and called for substantial debt relief by the year 2000. As an alliance of aid agencies, NGOs and groups such as the New Economics Foundation, the Campaign argued that debt was not only a modern form of slavery but also a symbolic reminder of how the fates of the North and South were bound up with one another. For every $1 sent as aid from the North over $10 returned to the North in the form of debt servicing. The Jubilee 2000 Campaign called for the 'chain of debt' to be broken and over 70,000 people gathered at the G8 Summit in Birmingham in May 1998 to demand further action. Some debt relief has been offered by the G8, but this is still considered inadequate. Without the removal of trade barriers and subsidy regimes in the North, the Global South will continue to be disadvantaged even if all forms of debt are cancelled. Ironically, the now third-biggest economy in the world, Germany, had its debts forgiven under the terms of the 1953 London Accord (although the Allied victors had earlier confiscated industrial materials and equipment).

---

institutions of globalisation are hardly new. For decades, people in the developing world have rioted when the austerity programs imposed on their countries proved to be too harsh, but their protests were largely unheard in the West. What is new is the wave of protests in the developed countries (Stiglitz 2002: 3).

Utilizing the Internet, email and the mobile phone, a new generation of anti-globalization movements has benefited from the global information revolution of previous decades. In other words, the anti-globalization movement is, despite its best claims, inherently global. Information and support networks bypass the mainstream media and in so doing create a fluid geography of resistance which is both local and global. Protest groups such as the People's Action Group and the Jubilee 2000 Campaign have been active in exposing high levels of corporate domination and/or the debt burdens endured by Third World states (see In focus 8.9).

Issues pertaining to North and South are bundled together in order to mobilize dissent against contemporary global capitalism. Information technologies give dissenting groups the capacity to co-ordinate high-profile protests against global institutions such as the WTO and organize worldwide days of action, as on 18 June 1999 (J18), when protests against globalization broke out in 100 cities across the world.

Sleepless in Seattle: the 1999 WTO meeting

One of the most high-profile acts of dissent to occur in recent years was in the midst of a WTO meeting in the city of Seattle in November/December 1999 (Fig. 8.4).

Better known for being the home of the Boeing aircraft company and the coffee chain *Starbucks*, as well as the setting of the popular American television show *Frasier*, Seattle

Figure 8.4 Resisting the WTO meeting in Seattle, November/December 1999
Photo: © Christopher J. Morris/CORBIS

became the scene of an extraordinary confrontation between American police and paramilitary forces (dressed in their chemical warfare outfits) and anti-globalization protestors. The third meeting of the WTO, otherwise known by anti-global critics as the 'star chamber for global capitalists' (St. Clair 1999: 81) was marred by accusations on the one hand of police brutality and on the other of mob violence. What is not in doubt is that observers such as Jeffrey St. Clair witnessed an extraordinarily diverse gathering of anti-globalization protestors. Members of Earth First!, trade-union movements, South Korean farmers and European students converged on Seattle's Convention Center, which hosted the 1999 WTO meeting. As one participant noted, 'The energy here is incredible. Black and white, labour and green, Americans, Europeans, Africans and Asians arm-in-arm. It's the most hopeful I've felt since the height of the civil rights movement' (cited in St. Clair 1999: 91). One of the 'guests of honour' was the French farmer José Bové, who addressed a crowd outside the local McDonald's restaurant. Shortly afterwards the building was stormed and destroyed. Bové was later seen handing out Roquefort cheese to his fellow protestors.

As part of their rejection of the WTO and the prevailing ideology of free-market economics, some of the protestors donned turtle outfits. While they were condemned by the mainstream media as 'mad' and 'deviant', the symbolism of the outfits was not lost on those hostile to the regulation of 'free trade'. Just before the Seattle meeting, the WTO had ruled that the US Endangered Species Act placed unfair restrictions on free trade because it insisted that shrimp could only be caught using turtle excluder devices. In other words, it was suggested that the 'turtle excluder devices' were (despite their conservationist motivation) an impediment to 'free trade'. Donning turtle outfits was thus meant to highlight the view of many anti-globalization protestors that the mantra of 'free trade' can be used to stymie attempts by others to conserve endangered species (see also McFarlane and Hay 2003).

The Seattle-based protests were not only significant insofar as they have stimulated further expressions of protest against world trade (in cities such as London, Doha and Genoa) but also illustrate that there are many people even in the United States who have profound misgivings about the character of contemporary globalization (Smith 2000, Wainwright, Prudham and Glassman 2000 and Fig. 8.5).

It is very tempting sometimes to assume that just because the United States is the largest economy in the world, ordinary Americans must consider globalization a

Figure 8.5 An anti-globalization 'fat cats' poster in Dublin
Photo: Klaus Dodds

widespread 'public good'. Increased economic competition and international flows of capital, commodities and people have unsettled many Americans (Steger 2003). Within Seattle and beyond, there is much concern as to the role the US plays as a geographical 'location' for overseas investment. As Agnew and Sharp (2002: 103) helpfully conclude:

> Both [sic] labor unions, environmental activists, and far right militia groups take exception to the idea that the United States and its regions are just locations for investment and disinvestments rather than parts of an abstract space of economic promise bequeathed to them by the frontier nation. This attitude is manifested in the increase in isolationist positions on both economic and military issues . . . Lurking within all of them [misgivings about WTO, UN and NATO] is the imperative to squeeze the genie of globalization unleashed by the frontier nation back into the territorial bottle of the United States.

The anti-globalization movement is an immensely diverse collection of people and ideas, and 'Seattle' demonstrated that thousands of citizens even in the United States are willing to take to the streets to protest. One should never underestimate the scale of dissenting voices and groups in the United States.

As befits the anti-globalization movement, arguments still rage over alternative world-visions to the WTO, the IMF and the US-led international trading and financial system. Some have suggested that a reformed and democratically accountable WTO might be able to impose a rules-based international trading system in a just and equitable manner. The WTO was never intended to promote human development per se and thus other global institutions might need to be created and/or reformed. Others contend that what is needed is a revolution on a global scale. George Monbiot, a British journalist and veteran protestor, has argued that alternatives to existing power structures have to be found (Monbiot 2002). A global democratically elected parliament is required, he contends, alongside reformed international institutions, which would help regulate international markets. In this new world, every state would be represented by one vote and thus unlike the current United Nations Security Council, the permanent veto of the United States and other Great Powers such as China and Russia would be ended. However, this still assumes that the nation-state remains at the heart of any reformed global system. Perhaps the real problem lies with the nature of the state.

The mantra of democracy, however well intentioned, is not a panacea, and the reform of institutions such as the WTO and IMF is not the same as a world government. The workings of global capitalism are perhaps too complicated for the introduction of a global democratic solution per se. As critics of Fukuyama (see Chapter 1) have contended, democracy is not universally accepted and Great Powers such as the United States have bypassed institutions such as the United Nations when they dislike proposed lines of action (such as the continuation of weapons inspections in Iraq). Moreover, for all their common misgivings about the IMF and the WTO, the anti-globalization movements remain as deeply concerned about national and local struggles for social and economic justice in Mexico, South Korea and Brazil (see Routledge 2003b).

Notwithstanding the different manifestos including democratic revolution and socialist planning, supporters and critics of the WTO and globalization alike recognize that the protests in Seattle, Cancun, Genoa and London are not just about trade and tariff protection; rather they are concerned with the future shape of global political and economic life. While many people took to the streets of Genoa, London and Seattle to protest, many more citizens did not. We have to be careful about simply focusing on publicized resistance. Other forms of 'hidden dissent' such as the illegal migrants stowing away on lorries, planes and trains rarely register in the anti-globalization debates, but this surely also calls into question how borders, territories and space are imagined and controlled. Ironically, the United States has long recognized that unrestrained markets are undesirable and has frequently used high tariffs and increases in defence spending to ensure that its economic fortunes are secured. Appointing Charlotte Beers as Under-Secretary of State for Diplomacy and Public Affairs in 2002, in an attempt to sell the 'American vision' of a free-trading world, will not be sufficient in itself to convince many anti-globalization activists that this vision of 'free trade' will ever be sufficiently attentive to social justice and fair access.

Conclusions

Paul Rogers, Professor of Peace Studies at Bradford University, has contended that the 1994 uprising in Chiapas and the various protests and acts of resistance in Chiapas, Seattle, London, Genoa and elsewhere form examples of 'prologue wars' (Rogers 2002: 78–80). While there have existed previous episodes of international solidarity such as during the Spanish Civil War in the 1930s and the anti-nuclear politics of the 1980s, these new struggles are indicative of future conflict. By way of contrast, the anti-colonial struggles ('epilogue wars') of the 1950s and 1960s are indicative of past trends. According to Rogers, the growing wealth–poverty divide, alongside contested access to resources such as oil and water, will help to define future conflicts. This perhaps underestimates the extraordinary and long-lasting legacy of colonialism, as the 2003 invasion and occupation of Iraq (a former British colony which has been bombed frequently since the 1920s) would imply.

Given the all-pervasive nature of globalization, it is unsurprising that protestors have directed their anger at symbols of global economic and political authority. Companies such as McDonald's, institutions like the WTO and European and American cities such as Genoa, London and Washington DC have all been targeted by dissidents and protestors. Anti-globalization activists, like global terror networks, can strike anywhere precisely because of the expansive geographies of global capitalism. As a consequence, even large economic powers such as the United States are drawn into defensive measures such as subsidizing domestic farming industries in order to protect the sector from global competition. The production and regulation of global political and economic space remains inherently divided and there are opportunities for dissidents and capitalists alike to continue to intervene.

Key questions

- Were anti-colonial resistance movements simply freedom fighters?
- Why did individuals and groups contest the binary politics of the Cold War?
- Why have so many people sought to resist contemporary global capitalism and the workings of institutions such as the World Trade Organization?
- What is anti-geopolitics and why is it important?
- Why do colonial representations (such as Orientalism) persist?
- Can globalization ever be truly fair and just?

Further reading

For an assessment of anti-geopolitics including Cold War dissidence see the essays contained in G. O Tuathail, S. Dalby and P. Routledge (eds.), *The Geopolitics Reader* (London, Routledge, 1998) and A. Blunt and J. Willis, *Dissident Geographies* (Harlow, Pearson Education, 2000). Geographical literature on 'Seattle' and dissent includes N. Smith, 'Global Seattle', *Society and Space* 18 (2000): 1–5, J. Wainwright, S. Prudham and J. Glassman, 'The battle in Seattle: microgeographies of resistance and the challenge of building alternative futures', *Society and Space* 18 (2000): 5–13 and T. McFarlane and I. Hay, 'The battle for Seattle: Protest, popular geopolitics and the Australian newspaper', *Political Geography* 22 (2003): 211–32. On anti-globalization see M. Steger, *Globalization: A Very Short Introduction* (Oxford, Oxford University Press, 2003), A. Cockburn, J. St. Clair and A. Sekula, *5 Days that Shook the World* (London, Verso, 2000) and K. Danaher and P. Burbach (eds.), *Global This!* (New York, Common Courage Press, 2000). P. Rogers, *Losing Control* (London, Pluto Press, 2002) is helpful in exploring how the struggles of the anti-globalization movements might be linked to contemporary discussions on security in the pre- and post-9/11 period. For the views of a high-profile British anti-globalization activist and journalist see G. Monbiot, *The Age of Consent: A Manifesto for a New World Order* (London, Flamingo, 2002) and for a critical review of globalization see, for example, M. Khor, *Rethinking Globalization* (London, Zed Books, 2001).

Websites

Earth First! www.earthfirst.org
Movement for Landless Workers www.mstbrazil.org
Jubilee 2000 Campaign www.jubilee200uk.org
For further information on anti-globalization movements see www.globalization.about.com

Chapter 9

GLOBALIZATION OF TERROR

Key issues

- How is terrorism defined?
- Why does a terror network such as Al-Qaeda challenge the presumptions of traditional geopolitics?
- What are the geopolitical objectives of Al-Qaeda?
- How did the George W. Bush administration respond to the threat posed by global terrorism?

American Airlines flight AA11 crashed into North Tower of New York's World Trade Center (WTC) at 8.35 a.m.

United Airlines flight UA175 flew into the South Tower of the WTC at 9.05 a.m.

American Airlines flight AA77 hit the Western side of the Pentagon at 9.39 a.m.

United Airlines flight UA93 crashed into a field near Stoney Creek in Pennsylvania at 10.03 a.m.

Despite its origins in the Cold War, the Al-Qaeda network, according to recent accounts, is the first multinational terror network of the twenty-first century (Gray 2002, Gunaratna 2002, Burke 2003 but see Halliday 2001). It has already succeeded in creating extraordinary expressions of anger and paranoia following the well planned and executed 11 September 2001 attacks on New York and Washington (with a fourth plane failing to reach its intended target, possibly the White House). The resulting fear is comparable to the worst excesses of the Cold War, when families on either side of the **Iron Curtain** were taught how to use a gas mask and instructed in making a backyard bomb shelter out of duct tape and plastic sheeting. Other analysts have turned their attention to the 1962 Cuban Missile Crisis (see Chapter 3) and explored parallels between the Kennedy and Bush administrations in their handling of chaos and confusion arising from threats to national security. At least Kennedy knew that Soviet missiles were being stationed on neighbouring Cuba. Unfortunately for the September 11th generation, the threat posed by Al-Qaeda is harder to identify in exclusively territorial terms.

In a manner reminiscent of the James Bond films of the 1960s, Al-Qaeda (Arabic for network and base or foundation) is the modern-day equivalent of SPECTRE, an organization based on loose cross-boundary networks and private finance rather than being state-sponsored and located in identifiable territory such as Iraq or Afghanistan.

This matters greatly because it means that the United States, as the principal victim of the September 11th attacks, has had to recognise that attacking one country (e.g. Afghanistan in December 2001 and Iraq in 2003) will not necessarily diminish the dangers posed by terror groups.

Why do terror networks challenge traditional geopolitics?

Terror groups such as Al-Qaeda, with their global networking, pose considerable conceptual challenges for students of geopolitics. Much of the geopolitics of the Cold War was dominated by states and their inter-state military alliances such as NATO and the Warsaw Pact, but the post-Cold War era appears more confusing. From the perspective of the United States, the Soviet Union (although equipped with a substantial arsenal) was territorially fixed. If a danger was posed by the Soviet Union, at least Cold War planners knew, via extensive intelligence gathering, where the military forces were accumulated. And vice versa. In the case of the Al-Qaeda terror network, this sense of 'security' is undermined.

The very nature of the network implies diffusion and the lack of a well-established territorial centre (see Castells 1996). Al-Qaeda's bases in Afghanistan and Sudan were temporary in nature and depended upon the support of existing regimes. Interestingly, when the United States mobilized its forces in the aftermath of September 11th, its response was to seek a *territorial solution*. It was suggested that, if the regime in Afghanistan were overthrown, the destruction of Al-Qaeda would automatically ensue. Unable (at the time of writing) to kill or capture Osama bin Laden, the United States and its allies have been frustrated by their inability to terminate this cross-territorial terror network. The 2003 assault on Iraq revealed no evidence of collusion with Al-Qaeda by Saddam Hussein's former secular regime. As we shall see, a rather traditional approach to geopolitics has prevailed and arguably this does not address the threats posed by global let alone regional terror networks.

Defining terrorism

The threat and deployment of terror is nothing new. The word 'terrorism' owes it origins to the English writer Edmund Burke, who witnessed at first hand the violent aftermath of the 1789 French Revolution and the overthrow of the French monarchy (Townsend 2002). Given the connotations attached to the word, the definition of terrorism remains extremely controversial.

Modern terror movements

While Al-Qaeda may have carried out a series of extraordinary atrocities, modern terror movements (often linked to the terror activities of revolutionary states or even

---

Definition box: Terrorism

Terrorism can be defined as the use of organized intimidation and or violence for the purpose of coercing a government and or community. Such an understanding of terrorism does not remove the troubling issue of *legitimacy*. Hence the frequently cited aphorism:

> One person's freedom fighter is another person's terrorist.

The second troubling definitional problem revolves around *agency*. Can a state act in a 'terrorist' manner? Is state-sponsored bombing deliberately targeted at civilian/ non-combatant populations and essential infrastructure such as water purification plants a form of terrorism? Is state-sponsored assassination of other heads of state a form of terrorism? When is violence inflicted by a state against another state or stateless peoples such as the Palestinians simply a form of self-defence? What makes terrorism different to regular military action? Is counter-insurgency a form of terrorism?

Terrorism implies, unlike regular military action, that the norms associated with international law and the Geneva Conventions have been ignored.

Finally our *understandings* of the meaning of 'terrorist' can change over time. The Zionist Stern Gang, which led resistance against the British mandate in Palestine in the 1920s to 1940s, was to become part of the political leadership of post-colonial Israel. Members of the Stern Gang had been denounced as 'terrorists'. Likewise Nelson Mandela, the first president of post-apartheid South Africa, had been denounced as a 'terrorist' and jailed for over 30 years by white minority regimes.

---

empires dating from centuries earlier) have been active since the late 1960s (see Chapter 8 for a discussion of anti-colonial violence and terrorism). Non-state terrorism was not invented by Al-Qaeda. Groups such as the Irish Republican Army (IRA), Baader-Meinhof, Shining Path, Ulster Defence Force (UDF), Red Brigade, Tamil Tigers, Jemaah Islamiyah, Euzkadi to Askatasuna (ETA), Popular Front for the Liberation of Palestine (PFLP) and Palestine Liberation Organization (PLO) loom large in any recent historical evaluation (see In focus 9.1). All of these groups have been adept at securing arms, intelligence and political support from either other terror groups or supporters located in other countries. To put it another way, the globalisation of communications and transport, especially since the 1960s, facilitated the development of modern terror groups.

Unlike the Al-Qaeda network, however, groups such as the PLO (seeking an independent Palestine) and the IRA remain driven in large part by a determination to secure a particular *territorial settlement* (a united Ireland in the case of the IRA). When terror attacks were deployed in Northern Ireland and mainland Britain, IRA operatives did not intentionally engage in suicidal terrorism as a form of martyrdom. Nevertheless, over 3,500 people were killed during the 30 years of the so-called Irish 'Troubles'. Despite the 1998 Good Friday Peace Agreement, people continue to die at the hands of

In focus 9.1: The Palestine Liberation
Organisation (PLO)

The Palestine Liberation Organisation (PLO) was primarily motivated by the 1948
occupation (called by Palestinians the *Nakbah* or Catastrophe) of historic Palestine
by Israel after previous defeats of the Egyptian, Jordanian and Syrian armed forces.
The PLO was formed in 1964 after an earlier association with the Al-Fatah movement. The PLO's main aim was (and remains) to see the creation of an independent
Palestine. Realizing that conventional military force was apparently hopeless against
the US-backed Israeli state, the PLO began a terror campaign against Israel and its
Western allies. Supported by the Soviet Union who had backed Israel's Arab neighbours in the past, the PLO virtually invented airline hijacking. Using a then recently
developed plastic explosive (Czech made) called Semtex, the PLO and another organization, the PFLP, hijacked airliners in the late 1960s and early 1970s. In September
1972, they also co-ordinated the high-profile kidnapping of Israeli athletes at the Munich
Olympic Games, which resulted in the death of Israelis and Palestinians alike after a
German rescue attempt.

The PLO under Yasser Arafat eventually emerged as a political force in the 1980s
and 1990s as renewed attempts were made to secure a peace settlement between the
Palestinians and Israelis. This has been problematic not only in terms of securing any
kind of settlement but also for most Israelis, who object that a former terror group
and its leader (as a long-term chairman of the Palestinian Authority) should now be
negotiating with the democratically elected Israeli government. Hence there was
serious discussion within the then Sharon government about either assassinating
Chairman Arafat or sending him into political exile, especially after 9/11 and the
declaration by the US of a 'war on terror'.

renegade Republican and Loyalist terror movements in Northern Ireland. In contrast,
the leadership of Al-Qaeda seeks to eradicate Western cultural, economic, military and
political influence from the Middle Eastern and Islamic world.

The rhetoric of Bin Laden suggests that he sees Al-Qaeda as part of a revolutionary vanguard embroiled in a 'clash of civilizations' with the West. Given the extended
remit of the Al-Qaeda network, the final part of this chapter explores how the US
government in particular has initiated a military and political campaign in response to
the September 11th attacks and the so-called 'war on terror'. The problem facing many
states and not just the United States is how to respond to a networked organization
that changes rapidly and unpredictably. Moreover, liberal democracies such as the UK
and the USA find it difficult to accept the erosion of civil rights entailed by repressive
legislation aimed at protecting the public from terrorist atrocities.

For students of geopolitics, the Al-Qaeda attacks on the United States reveal
extraordinary insights into how Osama bin Laden on the one hand, and the Bush administration on the other, have constructed particular representations of global space and
terror (Fig. 9.1).

Figure 9.1   Flag burning: protesting against 'imperial America'
Photo: PA Photos

These geopolitical representations are profoundly important in justifying and legitimating subsequent policy action as well as terror and counter-terror operations. This is clearly a delicate subject to raise in the aftermath of an event when nearly 3,000 people died in the most horrible manner. I believe that any fair assessment of the legacy of that dreadful day must entail a historical evaluation of the geopolitics of the Cold War. It helps to situate the September 11th attacks still further and avoids any temptation to see the event as without precedent or prior build-up.

Terrorism and the Cold War

Writing in the aftermath of September 11th, it is uncomfortable but necessary to consider how states and governments have played their part in facilitating modern terror movements. While much of the West's attention has been focused on the role of Libya, Iran and Syria in funding and supporting terror groups in the Middle East and beyond, it can be argued that the United States and the Soviet Union have also played a major part in shaping the Cold War and post-Cold War activities of terrorist organizations. During the 1970s and 1980s, for example, the IRA was supported by the fund-raising endeavours of Irish-Americans, and successive US administrations supplied arms to groups (condemned as terrorist by some governments and labelled resistance movements by the United States) in Nicaragua and Afghanistan, which later provided Osama bin Laden and the Al-Qaeda network with the means to transform their strategic objectives. Thus, and herein lies a major contradiction, governments including those of the United States, Russia, France and Britain have supported terror movements when it suited them but also condemned them when they turned against their national and/or strategic interests.

Types of terror

When reviewing earlier episodes of terrorism, it is important to acknowledge that terror groups such as the IRA, ETA and the Red Brigade have to be seen in conjunction with state-sponsored acts of terrorism. This means, as we have considered earlier, that two basic types of terrorism need to be recognised, since the mere word 'terror' conjures up thoughts of inhumanity, criminality and lack of political support (Townsend 2002).

First, there are the acts of terror carried out by states, often in response to terror movements but sometimes as part of a broader strategic policy directed against communism or the more recent 'war on terror'. The USA, for example, stands accused of refusing to condemn other states (client regimes such as Pakistan and Saudi Arabia and military regimes in Latin America; see Chapter 3) for perpetrating or supporting acts of terror against civilian populations around the world. Similarly, Britain has been accused of failing to criticize the excesses of other governments, especially the brutal Suharto dictatorship in Indonesia, which was supplied by British arms manufacturers (Phythian 2000). British jets were used on bombing raids carried out by the Indonesian air force against the people of East Timor, which had been illegally invaded in 1975. Was Britain, therefore, a sponsor of state terror?

Second, there is terrorism carried out by non-state groups such as the IRA, sponsored by other states such as Libya and or assisted by the fund-raising activities of the Irish-American-backed NORAID social movement. I suspect many people in the West tend to think of terrorism as an exclusively non-state activity.

State terror

Noam Chomsky, Professor of Linguistics at the Massachusetts Institute of Technology (MIT) has been one of the most strident critics of American foreign policy during and after the Cold War (for example, Chomsky 1991, 2001). One of Chomsky's most persistent accusations against successive American administrations is that they have often openly or covertly supported acts of terrorism against civilian populations throughout the Third World. Under the public banner of anti-communism and the promotion of freedom and democracy, the United States and others such as Britain and France stand accused of promoting activities that are the exact antithesis of their proclaimed values and public morals.

An example of a state supposedly protecting the interests of the 'free world' yet acting in contravention of such sentiments was the Kennedy administration (1960–3).

Four Third World leaders were targeted for assassination: Fidel Castro of Cuba, Rafael Trujillo of the Dominican Republic, Patrick Lumumba of Congo and Ngo Dinh Diem of South Vietnam. As the American journalist Seymour Hersh has noted, 'Jack Kennedy knew of and endorsed the CIA's assassination plotting against Lumumba and Trujillo before his inauguration on January 20, 1961' (Hersh 1998: 3). Lumumba, Trujillo and Ngo Dinh Diem were assassinated but despite their best efforts to implicate the American mafia and disaffected Cuban exiles based in Miami, the Kennedy administration failed to kill the socialist leader of Cuba, Fidel Castro. Is state-sponsored assassination a form of terrorism?

In their global struggle against the Soviet Union, the United States encouraged and or directed countless acts of terror against individuals and civilians more generally. In Vietnam, for example, biological weapons/defoliants such as Agent Orange were used in order to expose the military position of the VietCong. At the same time, however, the indiscriminate use of Agent Orange caused considerable suffering to civilians living within those environments. Alongside the use of biological warfare, US president Lyndon Johnson ordered the widespread conventional bombing of neighbouring Cambodia and Laos. More bombs were dropped on those two countries than on German cities and infrastructure throughout the Second World War (1940–5), with the estimated loss of 3 million lives. The Soviet Union also not only used terror against its own citizens but also was prepared to invade and terrorize neighbouring states such as Hungary (1956) and Czechoslovakia (1968) in order to protect socialist ideals. In practice, this meant that those who defied the wishes of Soviet leaders were dealt with ruthlessly.

Why do these sorts of activities matter now that the Cold War is over? Some critics of American and Soviet foreign policy have contended that the Cold War became a convenient excuse for legitimating acts of state-sponsored terror. American officials and presidents were prepared to countenance a range of policy options which ran contrary to American and international law, using the struggle against communism and the Soviet Union as an ever-ready justification for deploying all methods at its disposal. Likewise, Soviet activities were justified by reference to a global struggle against Western imperialists led by the United States. In the post-Cold War era, Russia remains embroiled in terror operations in the breakaway republic of Chechnya.

The United States and the Soviet Union were not the only proponents of state terror during the Cold War. Britain, France and Eastern European states such as Bulgaria all deployed terror-like activities in order to quell anti-colonial opposition, anti-communist forces and/or domestic dissent. Israel has been accused of behaving like a 'terror state' because of its policy of assassinating people it believes complicit in anti-Israeli terror operations. Apartheid South Africa (1948–90) has also been accused of behaving in a terror-like fashion when it bombed, invaded and occupied neighbouring 'frontline states' such as Namibia and Angola as well as executing assassination policies against anti-apartheid organizations such as the African National Congress (Hanlon 1986). Terrorism, therefore, whether directed against an individual or civilian society, has never been the exclusive preserve of non-state movements such as the IRA, the PLO and or the Red Brigade.

Non-state terror

In the case of Northern Ireland, the origins of the Republican terror movement, the IRA, lie with the contested territorial politics of Ireland as a whole. Notwithstanding a long history of colonial occupation dating from the eleventh century and the extensive use of terror by seventeenth-century English figures such as Oliver Cromwell, the 'Troubles' owe their existence to a sense of frustration and grievance felt by the minority Catholic community in Northern Ireland. By the late 1960s, increasing numbers of Catholics in the province were demanding equal civil rights in the face of widespread evidence that the Protestant majority was enjoying superior access to employment, housing and even voting. In a manner reminiscent of the Civil Rights movement in the United States, peaceful protests and civil disobedience became widespread, but Catholic communities actually supported the initial deployment of British troops for the sake of maintaining public order. Within five years, however, the presence of the British forces became more contested. Following atrocities such as the 1972 'Bloody Sunday' massacre of 13 civilians in Londonderry/Derry, the British army was seen as an 'occupying force' and an instrument of British state-sponsored terror. Protestant/ Loyalist terror groups such as Red Hand of Ulster, the Ulster Defence Force and the Ulster Volunteer Force were later to be accused of accessing British military intelligence as part of their plans to target Catholic communities and businesses.

The resurrection of the IRA was legitimized in the minds of many in Northern Ireland and elsewhere as a response to the terror-like activities of British troops. Britain was depicted as an imperial power which refused to 'decolonize' Ireland. The stated goal of the IRA was the removal of the British military and the promotion of a united Ireland. As part of its campaign against the British presence, the IRA bombed military and civilian targets in Northern Ireland and mainland Britain. From the 1970s onwards, bombings were carried out in cities and towns including Belfast, Birmingham, Manchester, Guildford and London. In December 1983 the world-famous shop Harrods was bombed and five people were left dead. In October 1984, Prime Minister Margaret Thatcher was almost assassinated as a massive bomb destroyed the Grand Hotel in Brighton during a Conservative Party Conference. In April 1992 the IRA targeted the City of London and severely damaged the Baltic Exchange. In August 1998 a splinter group of the IRA detonated a huge car bomb in the Northern Irish town of Omagh, killing 28 people.

While the mayhem caused by these bombings was indiscriminate, the IRA's targets were carefully chosen. By bombing mainland Britain the IRA deliberately ensured that the struggle for a united Ireland became part of the mainstream British political debate. It also had the effect of placing residents of London and certain Northern Irish cities such as Belfast and Londonderry/Derry in a perpetual state of apprehension. At the same time, the IRA and others have been accused of obtaining funds through extensive involvement in criminal activities such as prostitution, money laundering and extortion.

Non-state terror groups such as the IRA are thus important not only in the manner in which they challenge the legitimacy of nation-states but also in the way in which they can operate in (often covert) association with other countries. While the division between state and non-state terror has some heuristic value, these categories frequently become blurred.

Did terror operations achieve their stated political objectives?

In the case of the United States, the Cold War struggle against the Soviet Union was perceived as all-encompassing, so that any price (short of full-blown nuclear war) was seen as regrettable but necessary. One of the problems that bedevils analysis of state terror is definitional. In the case of Vietnam, for instance, the American intervention in the 1960s and 1970s remains highly controversial. However, ultimately many would claim that notwithstanding the attendant costs in human lives, the Soviet Union and its allies were prevented from realizing a global socialist revolution. Likewise it might be contended that the Soviet Union was motivated by a similar desire to prevent the triumph of American-led global capitalism. The costs surrounding the Cold War were truly staggering as the US spent over $10 trillion alone on defence during this period.

A resolution of the Northern Irish issue is still outstanding. Despite the financial assistance of Irish-Americans, the IRA has not been able to persuade the United States to impose a territorial settlement in its favour, partly because Britain is a close political ally of the USA and partly because Cold War priorities meant that aid and military assistance were directed towards the Middle East, South East Asia and Latin America. For its part, Britain has been unable to 'defeat' the IRA despite enlisting the Special Air Service (SAS) and the Secret Intelligence Services (SIS) and mounting covert operations including an alleged 'shoot to kill' policy against IRA members in the 1980s. In 1987, for example, the SAS killed three IRA operatives in the British colony of Gibraltar.

Arguably, however, the persistent terror operations of the IRA forced successive British governments to negotiate more closely with the Republic of Ireland government, and Anglo-Irish co-operation undoubtedly contributed to the implementation of the 1998 Good Friday Peace Agreement, which attempted to secure peace and security throughout the island of Ireland. Remarkably, the Republic of Ireland agreed to amend its constitution to remove its historic claim to Northern Ireland (the occupied Six Counties) in return for new governance involving closer cross-border co-operation. The IRA has agreed to demilitarize in return for Britain's reduction of its military presence in Northern Ireland. All sides have agreed, in principle, to pursue peaceful means and over time it is perfectly possible that a future referendum in Northern Ireland (on the question of the division between North and South) might lead to the creation of a united Ireland. The legacy left by terror and counter-terror operations continues to shape the aftermath of the 1998 peace settlement.

Other European countries too, such as Italy, France and Spain, have to contend with the activities of terror groups such as the Basque group ETA (demanding independence from France and Spain) and the Italian left-wing urban organization, Red

Brigade, whose political objective was to overthrow the capitalist state of Italy. Far-right groups in Italy were responsible for acts of terror such as the bombing of Bologna railway station in 1980.

Given the overwhelming scale of the Cold War conflicts, it is sometimes overlooked that the period between the 1960s and 1980s was routinely punctuated by struggles between states and terrorist organizations. The hijackings and assassinations carried out by the PLO, discussed above, provoked outrage around the world. Terrorism of whatever variety has long been an uncomfortable fact of life for citizens in many Western European states and Latin America. While Americans have occasionally been victims of terrorist atrocities committed in Europe and the Middle East, the United States was not really exposed to either domestic or international terrorism until February 1993, when Ramzi Ahmed Yousef, who was linked to Al-Qaeda, attacked the basement of New York's World Trade Center (WTC), killing six people and causing $300 million damage. Another major episode occurred in Oklahoma in 1995 (see In focus 9.2).

In terms of understanding both state and non-state terror operations prior to September 11th, it is important that we recognise that terror is not the sole preserve of large states such as the US or groups shaped or influenced by particular religions such as Christianity and Islam. Geographically, prior to September 11th, the US had been

---

In focus 9.2: Bombing heartland America: the 1995 Oklahoma bombing

On 19 April 1995, a bomb exploded outside a US federal office in downtown Oklahoma, killing 168 adults and children. After initial expressions of horror and disbelief that something of such murderous magnitude could have occurred in the US, pundits and politicians alike were swift to locate responsibility in the 'Middle East'. This region had long been represented as an outside space characterised by atavism, fanaticism, intolerance and turmoil. Arab-Americans complained that they were being subjected to harassment and violence as white America presumed collective guilt. Some aggrieved Americans even called for an all-out assault on Iraq and any other state in the region believed to be responsible for terrorism.

In June 1997, however, a white American man and 1991 Gulf War veteran, Timothy McVeigh, was found guilty of using the bomb as a weapon of mass destruction. The bombing had been deliberately timed to coincide with the US federal assault on the Branch Davidian compound in Waco, Texas, which McVeigh saw as an illustration of an overbearing federal government targeting minority white communities. Executed by lethal injection in 2000, McVeigh was, as the British geographer Matthew Sparke has noted, a 'Soldier-Patriot' who turned his murderous energies against the very nation-state that he was asked to fight for in the Persian Gulf. Awkward questions remain – such as how this particular form of white male paramilitary patriotism unleashed murderous levels of violence not only in the deserts of Iraq but also on the suburban streets of Oklahoma.

Further reading: *see* Sparke 1998: 198–223.

relatively insulated from state and non-state terror activities. Hence, many Americans had their sense of security profoundly shaken by the events of 11 September 2001.

Al-Qaeda and the geographies of terror and anti-terror

The Christian-era year 1979 was a dramatic year for Muslims. Four incidents were critical in shaping subsequent events:

1. The Soviets invaded Afghanistan.
2. Iran underwent a theocratic revolution.
3. Israel and Egypt brokered a peace deal.
4. A radical Wahhabi group occupied the Grand Mosque in the holy city of Mecca.

It has been contended retrospectively that the origins of the Al-Qaeda network owe a great deal to these events and the subsequent radicalization of the Islamic world (Burke 2003). The treatment of Palestinians, Bosnian Muslims and Saddam Hussein's Iraq in the 1980s and 1990s merely consolidated a view among many that the United States and others such as Israel and Serbia imperilled the global Muslim community (the Umma). The September 11th attacks on the United States occurred against a geopolitical backdrop, then, which had been shaped by incidents two decades earlier.

Trying to understand the geopolitical representations of Osama bin Laden and his associates requires us to explore these episodes in a little more detail. The Palestinian-Jordanian activist Abdullah Azzam, rather than the Yemeni Osama bin Laden, was the catalyst for the post-1979 creation of the Al-Qaeda network. Initial anger was directed against the communist Soviet Union (officially godless) and its desire to retain a client regime in strategically significant Afghanistan. Eventually the Soviet Union was forced to retreat in a manner reminiscent of the humiliating American withdrawal from Vietnam in the 1970s. American, Saudi and Pakistani support for the anti-Soviet resistance movements played a vital part in the defeat of the Soviet war machine. Saudi Arabia provided major financial support ($30 million in 1980 and $250 million in 1985) to encourage the creation of a Sunni Islamic regime in Afghanistan. Pakistan, under the military control of General Zia, conceived of Afghanistan as vital to Pakistani military strategy in the event of an Indian attack or even invasion if the Kashmiri dispute escalated dangerously. Religion and regional geopolitics were intertwined in a manner which was distinct from the global anti-communist strategy of the United States (see Sidaway 1998).

In the aftermath of the Cold War, the underlying strategic motivation for Al-Qaeda was broadened: to remove the United States and 'Western' political and cultural influence more generally from the Middle East and the Islamic world. With the dissolution of the Soviet Union in 1991, bin Laden declared war on the remaining superpower, the United States, in 1996. It was the United States and its regional allies such as Saudi Arabia and Pakistan that were now seen as the new 'Satan'. Unlike earlier European terror groups, Al-Qaeda alongside other 'Islamic' groups such as Hamas pioneered the routine use of suicide terrorism in the absence of sophisticated 'conventional' weapons held by the United States and other allies such as Israel. This predilection for suicide bombing was extended further in January 2002 when the first female operative, Waffa al-Edress, killed herself in Israel.

Why did the United States become the main adversary of Al-Qaeda in the 1990s?

Terrorism such as the September 11th attacks never occurs in a vacuum. The enmity towards the United States owes a great deal to the long-standing grievance over Palestine. Across the world, Muslims resent the continued injustice against the Palestinian people and the demonization of Islam. While many Muslims would deplore the terror activities of Al-Qaeda, the Palestinian question continues to agitate and aggravate. But Palestine is not the only source of resentment. Five interconnected geopolitical developments, which continue to have ramifications throughout the Middle Eastern and Islamic world, are widely cited by critics of the United States and its foreign and security policies.

First, since its creation in 1948 the state of Israel has received substantial amounts of financial and military assistance from the United States. It is also the region's only recognized nuclear power (although this has never been officially confirmed by Israel). As one of the most 'fundamentalist' nations in the world, Christian America (rather than just the often cited American-Jewish/Zionist lobby) stands accused of supporting Israel at the expense of Palestine.

Second, the US has developed close relations with pro-Western regimes such as Saudi Arabia, Pakistan and Egypt because of their resource potentials and or strategic locations. Thus American military and financial assistance has helped to sustain unpopular, corrupt and authoritarian pro-West governments. Israel and Egypt were two of the largest recipients of Cold War aid from the United States. The presence of American forces in Saudi Arabia (since Operation Desert Shield and Desert Storm in 1990–1) is considered to be especially offensive given their proximity to the holy sites of Medina and Mecca. The connection with Saudi Arabia is particularly significant given that 15 out of the 19 members of the September 11th suicide terror group were Saudi citizens.

Third, as a consequence of the Cold War, the United States and the Soviet Union were directly responsible for encouraging conflict and instability in Afghanistan, Horn of Africa, North Africa and the Persian Gulf states. The United States, despite supporting anti-Soviet forces in Afghanistan, was nonetheless resented by many Afghan citizens as a form of external interference. The Palestinians have, as many have contended, found themselves at the mercy of great power machinations before and after the Cold War. The United States remains the primary ally of Israel.

Fourth, the European Union did precious little to prevent Muslim communities such as those in Bosnia becoming victims of the armies of Serbia and Croatia. The massacre of Bosnian men and boys at Srebrenica in July 1995 (see Chapter 7) is still held to be indicative of this indifference. The Russians continue to be accused of 'terrorizing' the Muslim communities of Chechnya in the name of the 'war on terror'.

Fifth, the United States is despised because it is seen as the epitome of Western decadence. Many Islamic intellectuals and clerics believe that the West poses a mortal threat to traditional Islamic society and its codes of behaviour. Technologies such as the Internet are blamed for exposing Islamic societies to the corrupt political and cultural practices of the West.

All five factors have contributed to a view of the Islamic world as imperilled and of America as an imperial superpower. These points needed to be taken with some caution, however. In the case of Bosnia, for example, it could be argued that American intervention via NATO in 1995 actually helped to end further suffering of the Bosnian Muslim communities.

Al-Qaeda: a revolutionary vanguard?

Until the attacks on New York and Washington in September 2001, the Al-Qaeda network was known by a variety of names, including the World Islamic Front for the Jihad against the Jews and Crusaders. By the turn of the twenty-first century, the purpose of Al-Qaeda became to 'compete and challenge Western influence in the Muslim world' (Gunaratna 2002: 1). Al-Qaeda is a shadowy and virtual organization that was poorly understood by many Western intelligence agencies. With the benefit of academic and media accounts of Al-Qaeda by Rohan Gunaratna (2002) and journalists such as Jason Burke (2003), it is possible to trace some of the organizational and operational principles and strategies utilized by the terror network. Given its global networking, Al-Qaeda has demonstrated that there are no 'safe spaces' as even the heavily defended capital cities in the Western world are actual and/or potential targets. The British writer John Gray may be right, therefore, when he claims that:

> No cliché is more stupefying than that which describes Al Qaeda as a throwback to medieval times. It is a by-product of globalisation. Like the worldwide drug cartels and virtual business corporations that developed in the Nineties, it evolved at a time when financial deregulation had created vast pools of offshore wealth and organised crime had gone global. Its most distinctive form – projecting a privatised form of organised violence worldwide – was impossible in the past. (Gray 2003: 1–2)

Despite the fiery rhetoric against 'Jews and Crusaders', Al-Qaeda and its political leadership is, as Gray noted above, a thoroughly modern movement rather than a throwback to the European 'Middle Ages'. Its political leaders including Osama bin Laden are educated, often middle class and well travelled within and beyond the Middle Eastern and Islamic world. According to Jason Burke (2003), the network is loosely based and places such as Afghanistan and Sudan are simply useful in terms of providing training camps and co-ordinating centres. Although Al-Qaeda created its equivalent to the US School of the Americas in Afghanistan, the overthrow of the Taliban regime in December 2001 does not mean that Al-Qaeda has been fatally weakened: fund raising, training and other activities will simply engage other sympathetic regimes within the Islamic world. Bin Laden's extensive financial resources ensure a form of economic independence. The failure of the United States and the international community to adequately fund the rebuilding of Afghanistan's political and economic infrastructure means that Al-Qaeda may well still exist there, albeit in a reduced form.

Al-Qaeda has a basic ideology and 'way of seeing' which owes a great deal to the nineteenth-century notion of a revolutionary vanguard. As with nineteenth-century Russian revolutionaries, Al-Qaeda's founder Abdullah Azzam contends that:

> Every principle needs a vanguard to carry it forward and while focusing its way into society, puts up with heavy tasks and enormous sacrifices. There is no ideology, neither earthly nor heavenly, that does not require such a vanguard, that gives everything it possesses in order to achieve victory for this ideology. It carries the flag all along the sheer, endless and difficult path until it reaches its destination in the reality of life, since Allah has destined that it should make it and manifest itself. This vanguard constitutes Al-Qa'idah al-Sulbah [the Solid Base] for the expected society (cited in Gunaratna 2002: 3).

As the ideological father of Al-Qaeda, Assam is credited with establishing a network of operatives financed by Osama bin Laden's personal wealth, informal banking, the infiltration of the financial resources of Islamic charities, credit-card fraud in Europe and international donations. In the case of credit-card fraud, for example, it has been estimated that one Algerian cell in Britain raised over $200,000 in under six months in 1997. It is not thought, despite allegations made by American commentators, that Al-Qaeda has any involvement in the drugs trade.

While Sudan (1991–6) and Afghanistan (1996–2001) featured strongly in terms of the main centre of operations, Al-Qaeda continues to recruit new members in Sudan, Pakistan, China, Yemen, Philippines, Chechnya, Indonesia, Somalia and the Central Asian republics of the former Soviet Union. All these states are embroiled in conflict with 'Christian/Western' and or secular governments/opposition movements. According to some observers, close contacts have been maintained with Islamic terror groups/political movements and many have been invited to join the consultative council of Al-Qaeda, the Shura Majlis (Gunaratna 2002: 6–7). However, Britain, France (which is seen by many as supporting anti-Islamic regimes in Algeria and Morocco) and the United States have also proved to be recruiting grounds for Al-Qaeda. British Muslims, including those who worshipped in the Brixton Mosque in South London and the Finsbury Park Mosque in North London, travelled to Pakistan and Afghanistan but were later captured by the American forces who overthrew the Taliban regime in late 2001. Some British Muslims remain imprisoned by the United States at Camp Delta in Cuba and in Afghanistan.

The well-trained Al-Qaeda operatives carried out a series of outrages before the September 11th attacks. As with the attempted destruction of the WTC in 1993, the wide reach of the United States and its allies made it particularly vulnerable. In August 1998, the US embassies in Kenya and Tanzania were bombed and hundreds of Africans and 12 Americans perished. In 2000, the USS *Cole* was attacked in Aden harbour, and with the airborne attacks of September 2001 Al-Qaeda completed a campaign conducted on land, at sea and from the air. A second major attack by air was foiled when Richard Reid, a British citizen and Afghanistan veteran, was prevented from blowing up an American flight from Paris to Miami in December 2001. What all these attacks (and attempted ones) illustrate is the geographical extent of terror and counter-terror operations and hence the reason why American commentators (discussed in Chapter 1) feel the need to develop global analyses of the post-September 11th era.

In the period following September 11th, the Al-Qaeda network has been accused of carrying out a string of terror attacks across the Middle East, North Africa and South East Asia, including the Bali nightclub bombing in October 2002, which led to the loss of several hundred lives including over 80 Australians. This attack on the Indonesian islands which constitute the largest Muslim country in the world caused widespread panic among Western holidaymakers and was a crisis for President Abdurrahman Wahid, the head of a reforming civilian government. In May 2003, a series of suicide attacks were carried out in Casablanca (Morocco) and Chechnya, a bombing attack was launched against a foreigners' compound in the Saudi Arabian city of Riyadh and a number of co-ordinated assaults were perpetrated at Shell petrol stations in Karachi, Pakistan. The prime suspect in the fatal bombing of Riyadh was Khaled Jhani, a veteran of conflicts in Bosnia and Chechnya and an individual who had been trained by Al-Qaeda operatives in Afghanistan. According to Jason Burke (2003), Al-Qaeda operatives and/or local sympathizers carried out these assaults in places that stand accused of being pro-Western and publicly sympathetic to America's 'war on terror'.

The emerging geographies of counter-terror

Despite the fact that the Al-Qaeda network is widely recognised as a new kind of threat, the response of the British and American governments has been rather old-fashioned in the sense that they have deployed traditional (i.e. state-centred) forms of geopolitical reasoning and practice (Fig. 9.2).

The suicide bombings carried out on and after September 11th demonstrate the difficulty of responding to a group (rather than a state) that is prepared to send its operatives to their deaths. New legislation has been rushed through national parliaments in an attempt to circumvent terror-based activities. Rings of steel and or concrete have been constructed around public buildings such as the Houses of Parliament in central London. Security has apparently been improved at major airports such as London Heathrow and passengers are no longer allowed to carry nail scissors (though there are still glass bottles of duty free alcohol on board). The US and other Western governments have attempted to place stronger controls on terrorist financing, even if Al-Qaeda has proved adept at using banking services across the world. National airlines such as British Airways have been forced to temporarily suspend their flights to places such as Kenya which stand accused of harbouring Al-Qaeda activists. While the Global South suffers substantial losses in tourism, Northern states have sought to enhance security measures for their domestic citizens, but attendant dangers loom large with regard to

Figure 9.2 Britain and the United States: a special relationship between George W. Bush and Tony Blair?
Photo: PA Photos

the protection of civil liberties. The passing of the Anti-Terrorism Crime and Security Act in 2001 will place further restrictions on the movement and activities of British citizens. Arab-Americans and Asian-Britons complain of regular harassment and the people of Afghanistan and Iraq wait to see whether promises of UN and Western development assistance will be delivered to their societies over the longer term.

The 'war on terror' has also encouraged a general climate of 'counter-terror opportunism', as terror groups around the world are either labelled subsidiaries of Al-Qaeda or accused of collaborating with the network (Fig. 9.3).

Local struggles over territorial boundaries and sovereignty have been subsumed by a global discourse surrounding Al-Qaeda and the 'war on terror'. One immediate consequence of this development has been US reluctance to criticize its 'war on terror' allies, while at the same often ignoring the local reasons for territorially inspired terror groups. Kashmir, for example, remains a disputed territory between India and Pakistan but India has refused since 1947 to hold a referendum on its political future.

Afghanistan	3,000 suspected Al-Qaeda operatives held at Bagram Airbase
Australia	The government of John Howard, a prime supporter of the US assault on Afghanistan and Iraq, has introduced harsh anti-immigration policies and kept Afghan refugees on remote 'processing' islands such as Christmas
Britain	Over 400 arrests of Al-Qaeda operatives and new legislation such as 2001 Anti-Terrorism Crime and Security Act
China	Chinese authorities arrest over 400 ethnic Uighurs (Muslims) in Xinjiang Province
Cuba	600 held at Camp Delta, Guantanamo Bay
Georgia	US forces assisting Georgian government in assaults against Chechen rebels
India	2002 Prevention of Terrorism Act passed and further security operations in Kashmir
Israel	900 Palestinians held following security clampdown in 2002–3. 6,000 Palestinians held in Israeli jails. Israel's fight against Palestinian terror groups such as the Iran- and Syria-backed Hamas has now been rebranded as part of the general 'war on terror'
Russia	has intensified its 'anti-terror' operations against 'militants' operating in Chechnya
Spain	Increased police operations against Basque separatists
USA	Over 1,200 people arrested on suspicion of terrorist connections

Figure 9.3 The geographies of detention and deterrence
Data: Independent, 26 June 2003

New geographies of terror?

The new geographies of terror demonstrate only too clearly the inadequacy of Robert Kaplan's (1994) model of distinct 'tame' and 'wild' zones. Even if one accepts the terminology, 'wild' and 'tame' spaces are being increasingly intermixed as British cities such as London, Birmingham and Leicester find themselves embroiled in the networks of organized crime and terror. Rings of steel (or for that matter concrete) seem an inadequate defence against transnational crime involving people-smuggling, prostitution and money laundering. Terror operations involving suicide bombers are, as the Americans, Israelis and others have learned, extremely difficult to prevent within liberal democratic societies. In Israel, for example, suicide bus bombers attached to terror groups such as Hamas have dressed as ultra-orthodox Jews in order to disguise their murderous intent. Leaving aside the extraordinary human loss caused by the September 11th attacks, most of the victims of the recent terror campaigns have been located within the Global South and often within predominantly Muslim communities. The United States continues (at the time of writing) to imprison Al-Qaeda suspects in Cuba (Camp Delta at Guantanamo Bay) and Afghanistan (Bagram Airbase).

While only a small minority of the Muslim community across the world would condone the attacks carried out by the Al-Qaeda network, the end result has been to further demonize Islam and Muslims generally. It seems the challenge posed by Al-Qaeda for liberal democracies such as Britain and the United States is twofold: first, investigating and preventing further acts of terror, regardless of the scale of the proposed atrocities; second, acknowledging that many Muslims feel not only culturally vulnerable but also politically impotent in the face of continuing perceived injustices such as those surrounding the Palestine question and the rebuilding of Afghanistan. Developing a coherent multinational, multi-agency and multidimensional response to Al-Qaeda has proved to be extremely difficult, especially within the United States, which remains haunted by the losses incurred on 11 September 2001.

Responding to Al-Qaeda and the 'axis of evil': the Bush doctrine and regime change

How did President George W. Bush respond to the threat posed by the global terror network, Al-Qaeda? As mentioned earlier, his strategy was remarkably traditional in the sense that it singled out Afghanistan and Iraq. While Afghanistan provided a base for the Al-Qaeda network, the Iraqi regime had no known ties with such terror networks. Frustrated by their inability to capture or kill Osama bin Laden, America and other allied states stand accused of using the 'war on terror' as a convenient umbrella term for any conflict.

In the immediate aftermath of September 11th, most foreign governments and international observers were swift to express their sympathies with the plight of the United States. Fuelled by near-constant television coverage of the destruction of the World Trade Center, President Bush's popularity rose impressively as many sections of the American public rallied around the administration. Flag waving and patriotism abounded, so much so that friends and colleagues of this author who were in the United States at the time described the public reaction as similar to that in Britain to the death of Diana, Princess of Wales in August 1997. In the case of America, many citizens felt that the country's security, as well as their own peace of mind, was shattered.

In terms of forming a response, President Bush was accused of constructing a very simple binary opposition: *you're either with us or against us*. Unfortunately, this meant many Arab-Americans were summarily seized (in a disturbing echo of the internment of Japanese-Americans after the attack on Pearl Harbor, 7 December 1941) while others were to complain that they encountered new levels of hostility from strangers and were even evicted from flights after complaints from their fellow passengers that they looked 'suspicious'. One implication of September 11th was the internalization of danger as some American citizens felt that they were being linked to the threat posed by the Al-Qaeda network simply on the basis of their perceived racial and ethnic identities. At times of great distress and apparent danger, it is often the case that certain groups within societies are held to be either complicit in or even responsible for actions affecting a country. This is not a new phenomenon: during the Cold War, many people in the United States and Britain were accused of being crypto-communists (see Chapter 4).

Between September and December 2001, the US initially developed a broadly multilateral approach to the aftermath of September 11th. Bush was swift to visit the Islamic Center in Washington DC and to defend the pacific nature of the Islamic faith, and the administration sought to involve a wide range of partners to consolidate a response. The main base of the Al-Qaeda network was identified as being under the jurisdiction of the Taliban regime of Afghanistan. Unlike the Clinton administration, which in 1998 decided to bomb alleged Al-Qaeda bases in Afghanistan and Sudan, Bush proposed to move beyond the strategy of trying to geographically contain the network by actually invading Afghanistan and removing the Taliban regime. It was contended that with the removal of this regime or territorial hub, the Al-Qaeda network would be gravely disrupted – or possibly destroyed if its leader, Osama bin Laden, was killed in the process. Supported by the United Nations, the US-UK military campaign achieved the rapid collapse of the Taliban regime with the help of anti-Taliban factions (including the Northern Alliance) within Afghanistan. Unfortunately, the demise of the Taliban regime led to the further degradation of the country's infrastructure as anti-Taliban factions allegedly used American military power to settle old scores amongst themselves.

At this stage the US's self-declared 'war on terror' was centred on capturing or destroying the perceived perpetrators of September 11th. Within months of the Afghanistan campaign, the 'war on terror' was widened as so-called rogue or outlaw states were accused of dangerous liaisons with Al-Qaeda and or recklessly developing weapons of mass destruction. The State of the Union address by President Bush in January 2002 highlighted a change of military and political strategy, as he reimagined and re-represented global political space in a new and highly significant manner, designating some countries and regions as 'friendly' and others as 'dangerous'. For students of critical geopolitics, this speech provides a powerful example of practical geopolitics, illustrating how a US president responded to extraordinary events and new threats and dangers judged to be confronting America.

The axis of evil and the 2002 State of the Union address

In his State of the Union address, Bush revealed a rather more unilateral approach to September 11th. Three states – Iran, Iraq and North Korea – were identified as being part of an 'axis of evil'. They were charged with not only being indifferent to the current international legal order but also guilty (to varying degrees) of using force to settle disputes, developing weapons of mass destruction, violating human rights and harbouring terrorist groups such as Al-Qaeda.

Why use the term 'axis of evil'? The recently published memoirs of David Frum, a former presidential speechwriter, provide invaluable background information as to the origins of the phrase (Frum 2003). Recalling events surrounding the reaction of the Roosevelt administration following the Pearl Harbor attack in December 1941, Frum contends that Japanese and German fascism was akin to the beliefs and values of the governments of Iran and Iraq and the Al-Qaeda network. All stood accused of hating rational enquiry and free speech, and of promoting anti-Semitism and the celebration of murder (Frum 2003: 235–6). Originally Frum decided to call these latter-day terror states and groups an 'axis of hatred', but as the final draft of the State of the Union speech was being penned, the phrase was changed to 'axis of evil' and North Korea was added to the list because of its development of WMD. When Bush delivered the address, his railing was couched in these terms:

> States like these [Iran, Iraq and North Korea], and their terrorist allies, constitute an Axis of Evil, arming to threaten the peace of the world. By seeking weapons of mass destruction, these regimes pose a grave and growing danger. They could provide these arms to terrorists, giving them the means to match their hatred. They could attack our allies or attempt to blackmail the United States. In any of these cases, the price of indifference would be catastrophic. . . . The United States of America will not permit the world's most dangerous regimes to threaten us with the world's most destructive weapons (cited in Frum 2003: 239).

Why did this labelling of political space matter? Apart from the specific categorizing of particular places as part of an 'axis of evil', concern has been raised as to whether an entire region (the Middle East) is in danger of being condemned as 'dangerous' and or 'threatening' to the United States. This apparent 'guilt by geographical and or religious association' was a major preoccupation of the late Palestinian author Edward Said (see Chapter 8).

However, others such as the leading historian of the Middle East Bernard Lewis contend that there is a case to answer. In his influential account of the contemporary condition of the region, *What went Wrong? The Clash between Islam and Modernity in the Middle East*, Lewis contends that the Islamic world has failed to develop its full political, economic and cultural potential because of the existence of a 'culture of command' (Lewis 2001). The countries and cultures of the Middle Eastern and Islamic world are criticized by Lewis for their failure to promote individualism, democratic values and human rights. Little recognition is made of the contribution colonialism, the Cold War and Great Power intervention might have made to the prevailing geopolitics of the Middle East. As the British historian Francis Robinson has suggested, Lewis's conclusions appear to be a manifesto for widespread regime change. The last paragraph of the book contends:

> If the people of the Middle East continue on their present path, the suicide bomber may become a metaphor for the whole region, and there will be no escape from a downward spiral of hate and spite, rage and self-pity, poverty and oppression, culminating sooner or later in yet another alien domination; perhaps from a new Europe reverting to its old ways, perhaps from a resurgent Russia, perhaps from some new expanding superpower in the East. If they can abandon grievance and victimhood, settle their differences, and join their talents and energies, and resources in a common creative endeavour, then they can once again make the Middle East, in modern times as it was in antiquity and in the Middle Ages, a major center of civilisation. For the time being, the choice is their own (Lewis 2001 cited in Robinson 2001: 14).

Aside from Israel, the one country not mentioned by Lewis in his analysis of the contemporary condition of the Middle East is the United States. Judging by his conclusions, other states such as Britain and France must bear the responsibility for colonialism and external instability in the Middle East and Islamic world. While the book was published just before the invasion of Iraq in March 2003, its omission of the United States surprised many readers sympathetic to the plight of the Palestinians, for example.

Why do Lewis's arguments matter? His analysis has been judged to be highly influential in shaping neo-conservative intellectual opinion within the Bush administration (see Frum 2002, Frum and Pearle 2003). It has provided academic underpinning to foreign policy decision-making and thus could be considered a form of formal rather than popular geopolitics (see Chapter 4). The broader significance of Lewis's historical investigation should also not be underestimated for two reasons (even if he acknowledges the extraordinary cultural, scientific and political achievements of Muslim societies). First, he contends that Middle Eastern societies have failed to modernize and been unable to develop a secular civic society (with the exception of Turkey). The result has been a series of repressive tyrannies such as the Saddam Hussein regime in Iraq throughout the twentieth century. Israel, by way of contrast, has enjoyed democratic governance since 1948. Second, Lewis says comparatively little about European imperialism (the British for example created Kuwait in the 1890s and Iraq in the 1920s) and the impact of the Cold War. Interpretations of this type led neo-conservative intellectuals and officials within the Bush administration to conclude that so-called Islamic societies are incapable of embracing the values and practices of Western civilization (see In focus 9.3). Islam is, as Edward Said (1978, 1981) and others have warned,

---

In focus 9.5: Neo-conservative intellectuals and the Bush administration

Neo-conservative intellectuals such as David Frum, Richard Pearle, Condoleeza Rice and Paul Wolfowitz have been credited with providing an intellectual worldview for the Bush administration before and after September 11th. When President George W. Bush took office in January 2001, it was widely recognised that the son of the first President Bush (1989–93) was poorly travelled and very inexperienced in foreign affairs. His advisors, some of whom had served in his father's administration, helped to shape America's foreign and security agenda. This group of individuals is often credited with shaping a pro-Israel/anti-Palestine agenda and has been accused of enjoying close links with Christian fundamentalist and Zionist supporters. After September 11th, the neo-conservative intellectuals (often called neo-cons) fashioned President Bush's declaration on the 'war on terror' and provided an intellectual rationale for a pre-emptive doctrine which thus far has seen the United States attack the Taliban regime in Afghanistan and Saddam Hussein's regime in Iraq.

For further information on the neo-conservative worldview, see www.newamerican century.org.

simplistically equated with atavism and totalitarianism, ignoring the cultural and political diversity of the Islamic world.

This appraisal of the Middle East and the Islamic world underpins a worldview which was bolstered by military victory in Afghanistan in December 2001. The despatch of the Pakistan-backed Taliban regime encouraged President Bush to argue that in future, the United States will reserve the right to undertake military action – including pre-emptive action – as a matter of course. What is significant about the construction and implementation of the subsequently named '*Bush doctrine*' is the manner by which global political space is being constantly remapped and reinterpreted. For example, in early 2003 the White House released a paper on the National Strategy for Combating Terrorism, which divided the world into four types of states and their attitudes towards the 'war on terror': 'willing and able' states such as the European Union and Australia, 'reluctant' states such as Cuba, 'weak' states such as Pakistan, and 'unwilling' states such as Iran and Syria.

Geopolitical divisions: the role of Saudi Arabia

Notably, one country has never appeared on Bush's list of geopolitical concern: Saudi Arabia. Why might this be significant? If following the September 11th attacks, the United States judges the world to be a more dangerous place, it is of great interest to explore which countries are judged to be threatening and or supportive. While attention has been given to the alleged links between Iran, Iraq and Syria to Al-Qaeda, few have addressed the network's connections with the House of Saud. Bin Laden has had little known contact with the so-called 'axis of evil' states; indeed in 1991 he even tried to organize resistance to Saddam Hussein's Iraq. Within Saudi Arabia, Wahhabi Islam is considered unsympathetic to more tolerant or secular forms of Islam and the Saudi regime has sought to purge from the region alternative visions of Islam. With a long history of human-rights abuses and cultural suppression, the Saudi regime is arguably no better than Saddam Hussein's former regime in Iraq. However, the United States has always been prepared to militarily support this regime (witness the 1991 Persian Gulf War and the 60 years prior to that event) because of its considerable reserves of oil and its willingness for US troops to be based within its national territory. It could be argued, therefore, that Bush's mapping of global political space has been highly selective. Why were some states included in his 'axis of evil' while other possible contenders were excluded?

The diplomatic and military campaign against Iraq in 2002–3 demonstrated the readiness of the United States and its loyal supporters such as the British government of Tony Blair to disregard any state which attempted to veto their proposed course of action. Few would dispute that Saddam Hussein's tenure in Iraq was brutal. The 1988 chemical gas attack on Kurdish villages in northern Iraq was only one of innumerable atrocities carried out by a regime which used mass murder and torture as a matter of routine. The war against Iran in the 1980s also witnessed the use of chemical weapons against Iranians. However, as with events surrounding the 1991 Gulf

War, critics of the assault on Iraq remind us that throughout the 1980s the West had armed and supported this 'rogue state'. During that decade, Iran was seen as the ultimate fundamentalist threat and Iraq was often represented as a strategic bulwark in the region. After the Iraqi invasion of Kuwait in August 1990, Saddam Hussein was transformed from reliable ally to the devil incarnate. The representation of danger shifted decisively, and for critics this raises awkward questions about the erstwhile role of the United States and others in promoting and even sustaining Iraq's war machine and strategic ambitions.

Responding to terror:
the powerlessness of the United Nations?

The period between the passing of Resolution 1441 in November 2002 and the assault on Iraq in March 2003 severely dented the credibility of the United Nations. On the one hand, the Security Council found itself deeply divided about whether or not Iraq was complying with the terms of Resolution 1441. At the heart of the matter lay the issue of non-disclosure of WMD. The UN weapons inspection team under the leadership of the Danish diplomat Hans Blix could not find any evidence of widespread development of WMD. France and Russia, as permanent members of the Security Council, urged that Dr Blix and his team be given more time to ascertain the exact nature of Iraq's disarmament following earlier resolutions in the aftermath of the 1991 Gulf War. Opposing this course, the United States and the UK insisted that Iraq was still developing WMD and was capable of launching a potentially fatal attack 'within 45 minutes' against Western states and their allies. Moreover, Bush argued that the secular regime in Iraq was also striving to develop stronger co-operation with the Al-Qaeda network. When appeals to the UN Security Council to take their military intelligence seriously faltered, the US and the UK lost patience with the wider international community and launched a military assault on oil-rich Iraq in March 2003.

Within four weeks the regime of Saddam Hussein was destroyed as US-UK military forces overwhelmed a poorly equipped opponent. It must be recalled that even before the assault, Anglo-American aircraft patrolling no-fly zones had extensively bombed Iraqi military installations. Co-ordinated attacks from land, sea and air caused extensive damage to the infrastructure of Iraq, even if the resulting casualties were lower than in the 1991 Gulf War. It has been alleged that US military forces protected oil installations while hospitals, banks, municipal installations and national museums were left to the mercy of looters and opportunists. Since Iraq possesses the second-largest oil reserves in the world (Saudi Arabia has the largest reserves), it is not surprising that many critics continue to contend that the United States is more concerned with securing these sources of fuel than with preserving the legacies of the oldest civilizations in human history. Anti-war protestors concluded that WMD and Al-Qaeda provided a fig leaf for old-fashioned imperial ambition: 'oil-hungry America' simply 'colonized' Iraq. Little evidence has been found (as of this writing in 2004) either that

In focus 9.4: Iraq, Niger and the Internet

One of the key charges concerning Iraq's alleged development of WMD was that it had purchased uranium (so-called yellowcake) from the West African state of Niger. British intelligence officials believed that documents held in their possession confirmed the link with Niger. However, when the documents were handed to the International Atomic Energy Agency in Vienna for verification, officials used the Internet (and the Google search engine) to check on the veracity of the claims and found that the Niger officials cited in the document were either false or out of date. American intelligence officials distanced themselves from the claims surrounding Iraq and Niger, and at the time of writing, the British government has yet to provide further evidence publicly substantiating them. The IAEA confirmed that 500 tons of yellowcake were held at Iraq's nuclear research station at Tuwaitha. If Iraq was intent on developing a nuclear bomb, it would have needed to enrich the uranium (U235) rather than purchase more uranium ore (U238). Enrichment is expensive and difficult.

In 2002 the British government was forced to admit that, as part of its public dossier on the threat posed by Iraq, it had borrowed without acknowledgement Internet-based material by a former doctoral student who had explored Iraq's war machine under Saddam Hussein. Prime Minister Tony Blair remains accused of misleading the British Parliament and the electorate over the 2003 invasion of Iraq in a manner not dissimilar to Anthony Eden and his management of the Suez Crisis in 1956, when he was confronted with a strong nationalist Arab leader in Nasser.

Further reading: *see Independent* 2003.

Iraq was developing an extensive programme of WMD or that it was supporting the Al-Qaeda network (see In focus 9.4).

The war against Iraq in 2003 was justified by the US and its allies as a necessary response to the threat posed by Saddam Hussein's regime and its unwillingness to demonstrate that it had dismantled its WMD capabilities. The conflict highlighted the awkward role the United Nations often has to play, caught between the competing demands of member states. The schisms within the United Nations also underlined different geographical representations of danger, as the US emphasized the threat posed by Iraq (and its possible links to Al-Qaeda) while others pointed towards other states such as Saudi Arabia which may have been more complicit in the September 11th attacks. What we witnessed was a very important example of competing geopolitical visions.

Anti-geopolitics and the war against Iraq?

If we were looking for evidence of anti-geopolitics (see Chapter 8), the anti-war protestors in Europe, the Middle East and North America as well as former South African president Nelson Mandela provide some excellent source material. Powerful states such as the United States, Russia and China stand accused of ignoring international law and UN resolutions in pursuit of their own self-interested priorities. While the United States initially appealed to the UN Security Council, it mobilized substantial political and financial resources to marginalize and/or subjugate those who disagreed with its chosen course of action. Disturbingly for liberal democracies everywhere, the mainstream media in the United States devoted very little time and attention to anti-war protestors and groups. In May 2003 the actor Sean Penn paid the *New York Times* $125,000 to publish an essay which railed against President Bush and his foreign policy. Other actors such as Martin Sheen (the star of CBS's *West Wing*) who have expressed reservations about the Iraq campaign have been threatened with dismissal by television companies worried that American viewers and advertisers who supported the 2003 Iraq invasion would boycott channels that were associated with anti-war views.

The links between the military and the media in the aftermath of September 11th deserve careful consideration. Given the growing concentration of media ownership in the United States and the United Kingdom, a danger exists that alternative viewpoints are dismissed or marginalized in a climate where anti-war protestors are labelled 'anti-American', 'anti-British' or 'traitorous'. Public opinion was consistently supportive of the US-UK assault on Iraq and the vast majority of the media also supported the Bush and Blair strategy. Despite the diversity of media forms, fewer and fewer corporations and outlets are responsible for provision within the United States. Time Warner, for example, owns *Life*, *Time Magazine*, *Fortune*, CNN and Warner Bros Pictures, to name but a few. The high-profile media figure Rupert Murdoch owns Fox network, Fox News, *The Times*, the *New York Post* and the satellite channel DIREC TV. The Federal Communications Commission, headed by the son of Bush's secretary of state Colin Powell, is planning to relax ownership rules still further. However, large segments of the US are searching for international media sources such as BBC World News in the wake of fears that the critical quality of US media reporting is declining in the post-September 11th era.

Why do labels such as 'axis of evil' matter?

The labelling of certain global political spaces as an 'axis of evil' has profound implications not only for states such as Iran and North Korea (given the experience of Iraq in 2003) but also for the domestic political life of US citizens, especially Arab-Americans and African-Americans (see In focus 9.5).

The 'axis of evil' also neglects the role of other states such as Saudi Arabia and other political crises such as the Palestinian question. Following the 1979 Iranian Revolution and the occupation of the Grand Mosque in Mecca, the Saudi royal family was anxious to promote Wahhabi Islam in response to fears that Shia forms of Islam were on the ascendancy. State money combined with private donations was despatched to Afghanistan during the Soviet occupation in the 1980s, and this finance, along with Saudi political influence, was critical to the creation and evolution of the Al-Qaeda network. This has been extremely difficult for the Americans to acknowledge given

In focus 9.5: African-Americans and the 'war on terror'

In April 2003 a Pew Research Centre public opinion poll found that African-American support for the invasion of Iraq was the lowest of all groups surveyed (44 per cent), while white Americans overwhelmingly (77 per cent) endorsed the military action. Should we conclude that African-Americans are less patriotic? The distinguished African-American writer Maya Angelou offered another interpretation. While African-Americans shared the widespread sense of loss and dismay following September 11th, they were less inclined to support the view that the 'war on terror' was a new kind of threat. African-Americans have long contended with racist violence from white terror groups, and many would complain that everyday terror involving racism and violence does not receive the same amount of publicity as extraordinary terror such as the September 11th attacks. The United States was anxious to defend principles such as liberty and freedom during the Cold War, yet it appeared reluctant to extend these same basic rights to African-Americans until the 1960s onwards. Eventual improvements in democratic representation did not prevent the disenfranchisement of African-American voters who were not sent voting papers for the crucial 2000 presidential election in Florida. The United States and its white leaders remain, for many African-Americans, extremely selective when it comes to the application of democratic and liberal principles.

Despite their widespread scepticism about the war on Iraq, African-Americans provided the lion's share of the military labour. Over 37 per cent of the US military is composed of African-Americans. Two of the most high-profile members of the Bush administration which carried out the operations against Saddam Hussein are Secretary of State Powell and National Security Advisor Rice. Both are African-Americans.

Source: Younge 2003

their close relationship to Saudi Arabia stretching back to the 1940s. A Congressional Investigation into the September 11th attacks released in July 2003 was censored in one key area: the role of the Saudis in financing and supporting the Al-Qaeda network.

Within the Middle East, the publication of a UN/US/EU/Russian-sponsored 'road map' appeared to hold out the promise of some kind of settlement to the long-running territorial struggle between Israel and the Palestinians. Such a settlement would address one of the biggest perceived injustices within the Muslim world. The radicalization of young men and women across the Islamic world has undoubtedly been provoked and inflamed by the Palestine controversy. If a territorial settlement were secured, support for Al-Qaeda might diminish, notwithstanding bin Laden's talk of liberating the 'Muslim city of Jerusalem'. A final resolution to the Israel/Palestine dispute will have to involve the United States (given its support of Israel over the last 40 years) in order to ensure the viability of the two-state solution, but the Bush administration's willingness to act may be diminished by the presidential elections in November 2004.

Conclusions

One of the most important implications of September 11th for world politics has been the affirmation of the United States as the world's only hyper-power. The ending of the Cold War may have provided the global geopolitical context for this predominance, but the events of September 11th provided the catalyst not only for increased defence spending and military campaigning but also for an aggressive rearticulation of America's role in the world. Notwithstanding continued expressions of cultural trauma, the US still generates 30 per cent of the world's total economic product and spends $280 billion on defence (Hertsgaard 2003). The Bush doctrine implies that America is becoming more confident about its role in the world, and the schism with Europe over the 2003 Iraq war is reminiscent of earlier confrontations when America presented itself as an alternative to European civilisation. As with prior incarnations, George W. Bush's America presents itself as an agent of global discipline and universal values. However, as Michael Cox has noted, 'Winning one short war [in Afghanistan] is one thing; achieving a durable and acceptable international order after the guns have fallen silent is something else altogether' (Cox 2002: 264).

Is the greatest contemporary geopolitical challenge to American hegemony a terror network? The answer is probably no but it has shaken the self-perception of many American citizens as living in a place free from terror and war. The Al-Qaeda network points to a different kind of political world where states do not enjoy a monopoly of violence. It also uses the global movements of people, capital and commodities to its political and military advantage. Perhaps there is one simple if terrible truth following the September 11th attacks: globalisation makes terror available to everybody.

For students of critical geopolitics, the challenge posed by expressions of terrorism and counter-terrorism is immense. We have witnessed new practical geopolitical expressions of global political space (e.g. 'axis of evil') which have had far-reaching implications. New forms of inter-state co-operation have emerged; the proliferation and reach of terror groups such as Al-Qaeda means that the United States and Russia now co-operate very closely with one another in terms of intelligence sharing and terror prevention. The broader normative challenge facing states and their governments is to restructure the international political order (and its associated territorial dimensions) in a manner which promotes a more just and fair management and regulation of global political space.

Why do some people support or at least sympathize with Al-Qaeda? The United States and its European allies stand accused of being indifferent to the suffering of those in the Middle East and the 1.2 billion people who might be identified as Muslims. Terrorism never occurs in a geopolitical vacuum. As Jason Burke has wisely concluded, 'All terrorist violence, Islamic or otherwise, is contemptible. But just because we condemn it does not mean we should not strive to comprehend. We need to keep asking why' (2003: 250).

Key questions

- Can terrorism ever be a legitimate form of defence and/or resistance?
- Why did the September 11th attackers target the United States and the cities of New York and Washington DC?
- Why was the Al-Qaeda network able to operate in places such as Afghanistan and Sudan?
- Why did the United States label some states as part of an 'axis of evil'?
- Was the response of the United States in invading Afghanistan (2001) and Iraq (2003) disproportionate?
- What role should the United Nations play in combating global terrorism?

Further reading

On terrorism generally see C. Townsend, *Terrorism: A Very Short Introduction* (Oxford, Oxford University Press, 2002) and M. Buckley and R. Fawn (eds.), *Global Responses to Terrorism* (London, Routledge, 2003). On September 11th and the aftermath see F. Halliday, *Two hours that Shook the World* (London, Saqi Books, 2001) and K. Booth and T. Dunne, *Worlds in Collision* (Basingstoke, Palgrave, 2002). For a detailed analysis of the Al-Qaeda network see R. Gunaratna, *Inside Al-Qaeda* (London, Hurst and Company, 2002) and the first-hand account by the *Observer*'s chief Middle East correspondent, J. Burke, *Al-Qaeda: Casting a Shadow of Terror* (London, I. B. Tauris, 2003). J. Gray has written a provocative essay entitled *Al Qaeda and What it Means to be Modern* (London, Faber and Faber, 2002) and W. Blum, *Rogue State* (London, Zed, 2001) provides a very critical review of US interventions before 9/11. Y. Bodansky, *Bin Laden: The Man who Declared War on America* (New York, Forum, 2001) provides a highly stimulating account of bin Laden. For two excellent discussions of Islam see G. Kepel, *Jihad: The Trail of Political Islam* (Cambridge, MA, Harvard University Press, 2002) and J. Esposito, *What Everyone Needs to Know About Islam* (Oxford, Oxford University Press, 2002). Recent geographical work on 9/11 includes a 'Forum' in *Arab World Geographer* 4, 2 (2001): 77–103 and a special issue of *Geopolitics* 8, 3, (2003) entitled '11th September and its aftermath: the geopolitics of terror'. For a post-colonial analysis of September 11th and the subsequent 'war on terror' see D. Gregory, *The Colonial Past* (Oxford, Blackwell, 2004). For further information on human rights protection in the post-9/11 era see W. Schulz, *Tainted Legacy* (New York, Nation Books, 2003) and D. Cole, *Enemy Aliens* (New York, New Press, 2003).

Websites

Department of Homeland Security (USA) www.dhs.gov
Centre for the Study of Terrorism and Political Violence www.st-andrews.ac.uk
Contemporary media reports http//warincontext.org
Stockholm Peace Research Institute www.sipri.se

Chapter 10

CONCLUSIONS

The repercussions of September 11th have profoundly reinforced the continued importance of the nation-state, territory and the politics of identity. It has also provoked renewed interest in empire and imperial behaviour with specific reference to the George W. Bush administration and the occupation of Iraq in 2003 (see Hardt and Negri 2000, Flint 2003, Hyndman 2003, Ferguson 2004). Critical geopolitics and allied disciplines such as International Relations and Ethics have quite rightly investigated how geographical expressions such as 'axis of evil' have reinforced representations of the United States as not only a 'vulnerable' state but also a morally righteous imperial state (see for example Singer 2004). This has been important in vindicating and legitimating US military intervention in Afghanistan and Iraq as well as placing further restrictions on international air travel and civil liberties within the United States. Likewise, those who oppose, for example, the 2003 invasion of Iraq define their responses as inherently directed against an identifiably territorial presence: the United States. External enemies and dangers continue to play an important role in reinforcing domestic identities and politics within states. The decision by Spanish voters to elect a new government in March 2004 reflected an overwhelming rejection of Prime Minister Aznar's close association with the American-sponsored 'war on terror' following terrorist attacks on Madrid's railway stations that were initially blamed on the Basque separatist group ETA.

In an era often characterized as one of intensive globalization, the 'war on terror' has not only witnessed the resurrection of Cold War-like competition between the United States and Russia but perversely also stimulated increased co-operation between states. While Russia and America publicly agree on the need to combat terror, they have both sought to increase their military presence in resource-rich Central Asia. In Kyrgyzstan, for example, the Americans established a new airbase at Manas following the 2001 assault on Afghanistan. Fearful of a possible extension of American influence in Central Asia, the Russians established a new base at Kant. The two bases are separated by 55 km of semi-desert. President Putin of Russia justified the decision on the basis of combating the activities of terror groups inside and outside the Russian Federation. These activities have helped to stimulate further debate in Central Asia and elsewhere about the strategic importance of this apparent 'pivot region' (originally Halford Mackinder's 1904 term). Formal, practical and popular forms of geopolitical reasoning are in much abundance.

The significance of September 11th should not be seen in isolation, however. Only the most parochial observer would contend that global politics in the contemporary era is defined above all by responses to that terrible day. It is probably still far too early to make a long-term assessment of the repercussions. Any attempt to do so would have to take account of the reaction of states and governments. As we have explored in Chapter 9, the ongoing 'war on terror' points to the continued importance of geographical scale (global, regional, national and local) and the extra-territorial practices of the powerful states including Russia, the United States and the United Kingdom. Geographical descriptions of political space (such as 'axis of evil' or 'pivot region') need to be carefully analysed and contested rather than simply naturalized.

Similarly, it would be unwise to claim that other less spectacular events and processes associated with 24/7 news reporting, international financial markets and border penetration affect citizens, groupings and states equally. The world is also a highly unequal environment in the sense that some states, such as the USA, Japan and China, enjoy far greater influence over global events and processes (including new trade regulations concerning investment and commerce) than entire regions such as Sub-Saharan Africa and Central America. Unrestricted movement of people and capital remains a comparative luxury for those who are located in the Global South. For the refugee or illegal immigrant, borders can remain hostile obstacles in their quest for sanctuary or a better life. A few feet can seem like a matter of life or death.

Our theorizing therefore needs to be sensitive to the very real divisions and inequalities in the world. As Archibugi reminds us, 'In the era of the computer, a third of the inhabitants of our planet have never used a telephone; cosmopolitanism remains the prerogative of an elite' (2002: 26).

For students of geopolitics, the significance of an increasingly interconnected if highly unequal world is multifaceted:

- Events in one part of the world may have profound implications for other parts of the world, regardless of distance and territory.
- The control of territorial space has become more problematic as flows of people, terror, money and ideas challenge the ability of states to control such movement. Ideas of national sovereignty and territorial jurisdiction are becoming increasingly problematic.
- Globalization is producing a more unequal and hierarchical world in which some states are better equipped than others to take advantage of trade opportunities and/or resist economic recession.
- The distinction between the 'global'/'local', 'inside'/'outside' becomes harder to sustain in the light of a range of transnational flows and networks including terror.
- Territorial borders, as a consequence, do not delimit political authority and community.

In this book I have tried to demonstrate how issues such as nuclear proliferation, humanitarianism and financial and informational flows can sustain particular representations of a 'borderless world'. The dangers posed by a nuclear war, for example, clearly have implications for all of us. But the particular geographies of nuclear testing were such that specific communities suffered and continue to suffer in spite of the absence of a global nuclear conflict. Appeals to global humanitarianism have placed new pressures on the nation-state and the claims of governments to enjoy absolute sovereignty over particular territories. However, it is also apparent that the United Nations and other international agencies have been prepared to react more strongly to some human-rights abuses (for example, in Kosovo) than to others such as in Angola, Mozambique and areas in the Russian Federation (Harriss 1995, Hoogvelt 1997).

Finally, the preceding chapters have been informed by a conviction that geopolitics, like other intellectual fields, cannot hide behind a veil of objectivity and neutrality. Knowledge is always partial, and theories – whether they be realist or inspired by the critical geopolitical literature – are never divorced from political, cultural and/or racial agendas (Agnew 2002: 180). This is a very important shift from earlier work in geopolitics, which was premised on the view that the academic researcher could investigate, define and classify the world in a disinterested and disembodied manner. Hence, my reflections on global geopolitics should be seen for what they are: partial, situated and unquestionably contestable. I have no doubt this book would and should look very different if written by an academic based in Abuja, Tehran or Singapore. The task of critical geopolitics is not only to contest dominant ways of seeing and knowing but also to contribute both in the academy and elsewhere to alternative forms of analysis and dissent.

Students of geopolitics should retain a sense of humanity, justice and commitment for those oppressed, tortured and deprived of basic human or community rights. While the development of an ethically literate form of critical geopolitics remains a challenging project, we can begin by insisting that places are human constructions and that the moral dilemmas posed by the appalling conditions of the Congo are also geographical dilemmas. How are places defined? How do we respond to the suffering of others in distant and not so distant places? Why do we value some places more highly than others? How do we resist imperial and/or colonial representations of place and people? A question I feel sure that many Africans must ask routinely when they witness their low profile of Africa in the United Nations. Places are not simply backdrops to human life. They help construct and shape our relations with others. In the Euro-American world, geographers have a privilege that is denied to many students and scholars working in the Congo, the former Yugoslavia, Myanmar (Burma), Iraq or East Timor: in the words of the French philosopher and historian Michel Foucault, they have the ability 'to speak truth to the face of power'. Critical geopolitics needs to continue to make a difference through our intellectual commitments and normative engagements with the world around us.

Further Reading

On various themes mentioned in the Conclusion: C. Weber, 'Flying planes can be dangerous', *Millennium* 31 (2002): 129–48. L. Ling, *Postcolonial International Relations* (Basingstoke, Palgrave, 2001). D. Archibugi 'Demos and cosmopolis' *New Left Review* 13 (2002): 24–40.

GLOSSARY

This short glossary is intended to offer some definitions concerning key terms such as 'geopolitics', 'neo-liberalism' and 'new world order'. These entries should be regarded as suggestive rather than definitive. For further details on key geopolitical terms see *The Dictionary of Human Geography* (edited by R. Johnston, D. Gregory and D. Smith, fourth edition, Oxford Blackwell, 2000); G. O Tuathail, S. Dalby and P. Routledge, *The Geopolitics Reader* (London: Routledge, 1998) and the older but still very useful *The Dictionary of Geopolitics* (edited by John O Loughlin, Washington: Greenwood Press, 1994).

axis of evil: A term used by President George W. Bush in his 2002 State of the Union Address to describe three states: Iran, Iraq and North Korea. Although there was little evidence to suggest that these three states collaborated with one another in terms of the promotion of terrorism, they nonetheless stood accused of funding and supporting terror-based activities in places such as Palestine. North Korea was included in the list because it has been reluctant to allow international inspections of its nuclear facilities.

Berlin Wall: The wall constructed by the Soviets to divide the city of Berlin in Germany into two sectors in 1961. After the Second World War, Berlin was occupied by four powers and their administrative sectors: France, the UK, the USA and the Soviet Union. After a series of crises, the Soviets decided to build a concrete wall across the centre of the city. It became a powerful illustration of a divided city and a European continent split between the capitalist West and the communist East. In November 1989, Germans toppled the East German regime and dismantled parts of the wall. Germany was reunified in 1990 and the capital has been moved from Bonn to Berlin.

Cold War: a term invented by the US journalist Walter Lippman to refer to the conflict and tension between the USA and the Soviet Union (called the superpowers because of their military strength) between 1945 and 1991. The term 'Cold War' was popular because it implied a 'frosty' relationship between the two countries rather than outright war. However, the implications for the countries in the Third World were dramatic as superpower rivalry either worsened civil wars or provoked new conflicts. Europe did not escape either from the violence of the Cold War as rebellions against Soviet rule in Hungary (1956) and Czechoslovakia (1968) were ruthlessly crushed. The ending of the Cold War is usually dated from the fall of the Berlin Wall in November 1989 and the subsequent dismantling of the Soviet Union in 1991.

communism: This refers to a political theory or form of government in which all forms of property are publicly owned and where the central state plays a major role in shaping social, economic and political relations. In contrast to liberal democracies, the individual under communism usually has restricted rights relating to voting, freedom of movement and property ownership. During the Cold War, communist regimes in the Soviet Union and Eastern Europe were often ruthless in dealing with any citizen or group of citizens who defied the authority of the official communist parties. High-profile dissidents were either murdered or sent to brutal labour camps in Soviet Siberia called Gulags.

critical geopolitics: This term refers to a recent body of literature in North America and Europe which explored the geographical assumptions and understandings underpinning foreign policy making and theories of world politics. Particular attention has been given to the use of geographical metaphors (e.g. heartland, containment, 'domino effect' during the Cold War) and their significance in popular and formal geopolitics. Critical geopolitics has demonstrated that geopolitical themes are to be found in the cinema, newspapers, television and in music. In contrast to **geopolitics**, explanations are sought to determine how geographical labels and designations enter into popular and formal discourse rather than to imply a strong causal relationship between global physical geography and state behaviour.

discourse: The persistent assumptions, claims and modes of analysis that both enable and constrain debate and practice.

formal geopolitics: The body of literature written by academic specialists on geopolitics, including classic writers such as

Halford Mackinder and more contemporary scholars such as Colin Gray.

gender: This refers to the assumptions placed upon and the divisions made between men and women. It is not the same as 'sex', which refers solely to the biological differences between men and women. Recent work in gender and world politics has revealed that economic and political restructuring of the world economy has very different implications for men and women, not least because women are often expected to care for children while at the same trying to earn money through work.

geopolitics: Originally coined at the turn of the twentieth century, this referred to a particular approach to world politics which stressed the significance of territory and resources. During the post-war period, Anglo-American geographers were reluctant to use the term 'geopolitics' because they felt that it had inspired Nazi Germany's policies of spatial expansionism. In the 1970s, however, political figures such as Henry Kissinger and Ronald Reagan used geopolitical language to describe international affairs and the Cold War against the Soviet Union.

globalization: A term widely used in the social sciences to point to the intensification and geographical spread of international interaction. Academic ideas of globalization include concepts such as interdependence, internationalization, interpenetration, modernization, time–space compression, universalism and integration. From a geographical perspective, the literature on globalization raises profound challenges, not least because some authors have argued that territorial space has become less important in shaping world affairs. The new millennium, according to some writers, would be characterized by the domination of global capitalism itself sustained by endless images of a borderless world, virtual financial flows and the complete domination of physical space. However, these varied representations of globalization rest on a series of assumptions about the social and political world which have yet to be determined.

hegemony: The capacity of a particular political or cultural group to exercise control and perpetuate inequality through the deployment of particular ideas and practices rather than through the use of force. Employing the ideas of the Italian thinker Antonio Gramsci (who was imprisoned between 1928 and 1935), many scholars have explored how citizens might actually support ideas and practices which either curtail their liberties or impose restrictions on others. Hegemony implies, therefore, more than just the dominant ideology of elites and **popular geopolitics** which seeks to explore how ideas about global geopolitical space are embedded in everyday life via the education system and media outlets.

intergovernmental organizations: Usually composed of groups of states which have created a governing body for the management of an aspect of international affairs. They include military bodies such as the North Atlantic Treaty Organization (NATO) created in 1949 by 12 states for the purpose of co-ordinating the defence of the North Atlantic against Soviet forces during the Cold War. Far from dissolving with the Cold War, NATO has assumed considerable influence in the post-Cold War world and now plays a significant role in supporting UN operations, as witnessed in Bosnia. In May 2004, NATO's membership was expanded again to include Bulgaria, Estonia, Latvia, Lithuania, Romania, Slovakia and Slovenia, and this consolidated an earlier

round of expansion in 1999 involving the Czech Republic, Hungary and Poland.

Other IGOs include the International Labour Organization of the United Nations. Created in 1919, it was formally adopted by the UN in December 1946. The ILO seeks to improve the working conditions and to protect the rights of workers around the world in terms of health, safety and trade-union membership. In 1969 the ILO was awarded the Nobel Peace Prize in recognition of its contribution.

internationalization: The enhanced interaction between and within states as a consequence of transboundary processes such as trade, investment, war, migration, environmental issues and/or crime. Recently, analysts have preferred the term 'globalization' in order to emphasize that states are not the only political actors to be influenced by these trends.

International Monetary Fund: Established by the Bretton Woods Agreement in 1944 and operating since 1947 under the United Nations, the IMF advises governments and the World Bank on fiscal policies such as taxation, interest-rate policy and the funding of public policy programmes.

Iron Curtain: A term first used by British prime minister Winston Churchill in 1946 to describe political and geographical divisions within continental Europe. It rapidly became a shorthand term in the West during the Cold War to describe the communist regimes of Eastern Europe and the Soviet Union.

neo-colonialism: A mode of economic and political control exercised by powerful states (usually former imperial powers) and world cities such as London and New York for the subjugation of the Global South. While the dominated states have secured formal independence, they nonetheless remain embedded in relations of exploitation and dominance. Hence the claim that colonialism remains at the heart of the collective predicament of the Global South.

neo-liberalism: A corpus of ideas and theories which became extremely important in the 1980s and 1990s. The free market (as opposed to state-sponsored communism) is seen as the most desirable mechanism for ensuring economic, social and political life. While its intellectual origins lie in the eighteenth-century writings of people such as Adam Smith, neo-liberalism has become strongly associated with the United States and the worldview of institutions such as the WTO, IMF and World Bank. All these organizations believe that the market can ensure the best distribution of goods, ideas and knowledge. Thus, the state and associated society should facilitate rather than actively intervene in the workings of the market.

new world order: A phrase popularized (not invented) by US president George Bush Senior in August 1990 to describe the world at the end of the Cold War. It was hoped that this would be a moment for unprecedented international co-operation and a new opportunity for the United Nations to help govern a more peaceful world. Unfortunately, humanitarian crises in Iraq, Somalia, Bosnia and other parts of the world led some commentators to talk about a 'new world disorder'.

Non-Alignment Movement: Created in 1961, this Southern political grouping was committed to five major principles: peace and disarmament, economic justice, self-determination, cultural respect and multilaterial co-operation within bodies such as the United Nations. The political glue

which bound NAM together was derived from a common desire to negotiate their own governance within the context of a superpower struggle and decolonization.

non-governmental organizations: These are political organizations which operate independently of states and other organizations such as the WTO. They often seek to pursue radical political agendas independently from the formal realm of politics due to a belief that their goals should not co-opted by mainstream politics. In terms of organizational structure, many NGOs are composed of flexible networks rather than rigid hierarchies.

North and South: These terms became increasingly popular with political observers in the 1980s. Many people in the Third World argued that with the ending of the Cold War, the most fundamental differences were due to inequalities of wealth. The terms 'North' and 'South' are geographical in the sense that the wealthiest countries of the world tend to be located in the Northern and the poorest in the Southern hemisphere.

Pearl Harbor: A reference to both a historical event and a mode of explanation, which became popular following the September 11th attacks on the United States. As an event, Pearl Harbor refers to the Japanese attack on US naval forces stationed at Pearl Harbor in Hawaii on 7 December 1941. The assault led to the US declaring war on Japan and thus entering into the Second World War. The September 11th attacks were thus the first time since Pearl Harbor that US territory had been subjected to 'external attack'. As a mode of explanation, 'Pearl Harbor' became a shorthand term to refer to an unprovoked attack in which complete responsibility unquestionably lay with the aggressor and thus any response by the United States was legally and morally justifiable.

political realism: A body of thought which gives great emphasis to the role of the state in shaping international politics and contends that the state's greatest political priority is national security. It also argues that in the absence of an all-powerful world government (not the United Nations), states have to operate in an international arena characterized by anarchy.

popular geopolitics: A term used in **critical geopolitics** to refer to the geographical representations of global political space found in popular cultural forms such as cartoons, films, novels and music. It is argued that popular culture plays a significant role in reproducing certain hegemonic values about political space, which need to be carefully examined.

post-colonialism: This refers to a body of writing which seeks to challenge Western ways of knowing the world. It is always anti-colonial and frequently concerns itself with the experiences of marginal and/or colonial peoples rather than addressing the interests and perspectives of the colonial powers.

practical geopolitics: This refers to the everyday forms of practical reasoning used by political leaders and civil servants to explain and justify their foreign and security policies.

structural adjustment programmes: These programmes were designed in the 1980s and 1990s by the IMF and the World Bank in an attempt to pressurize indebted Southern states such as Argentina, Zimbabwe and Thailand to undertake public-sector savings in return for further loans and grants. In effect, SAPs often forced these governments to cut their

spending on health care, education and other public sectors in order to achieve spending cuts.

superpowers: The term was coined during the Cold War to reflect the political and military significance of the United States and the Soviet Union. Since the demise of the Soviet Union in 1989–90, the United States has been described as the sole remaining superpower or hyperpower.

Third World: A term invented by French social scientists in the 1950s to describe the continents of Africa, Asia, Latin America and Oceania. These parts of the world were also known as the 'developing world' because they were considered to be distinct from the advanced economies of the First World (sometimes called the North and West) and the Second World of socialist states (sometimes called the East or communist bloc). Many writers prefer the term 'South' because they think that the term 'Third World' is derogatory to peoples living in the Southern hemisphere.

traditional geopolitics: A body of writing which places emphasis on the importance of geographical factors such as physical location in shaping the behaviour of states and governments. Some of its underlying assumptions are similar to those of political realism. Contributors to this genre of literature are often eager to offer policy-related advice to governments and political leaders.

transboundary: The term implies processes and phenomena such as acid rain and money which (by their very nature) have a capacity to cross territorial and other administrative boundaries.

transnational corporations: These are large organizations which operate in a number of different national economies. They are by definition **transboundary** because of their often complex networks of locations and activities. It is estimated that around 30 per cent of the world's trade is conducted between transnational corporations and this growing influence has been assisted by the development of communication and financial networks which allow money and trade to flow around the globe at an ever faster rate.

UNESCO: The United Nations Education, Scientific and Cultural Organization was created in November 1946 for the specific purpose of promoting collaboration amongst nations through education, scientific, media and cultural projects. It has been at the forefront of promoting the free flow of information, improving educational opportunities in the South, maintaining a United Nations University in Tokyo, safeguarding places and sites of great ecological, cultural and historical importance and raising the literacy rates of women. The latter has been considered a crucial dimension in the various UN-sponsored programmes for improving the lives and conditions of women in the North and South. It has been a controversial body at times because of its calls for fundamental reform in the ways in which information is collected, exchanged and disseminated free of censorship and state interference.

war on terror: A term used by President George W. Bush to describe his post-September 11th strategy. He committed the full military and diplomatic strength of the United States to combat the threat of terrorism regardless of geographical location. See other entries such as '**axis of evil**'.

World Bank: Created by the Bretton Woods Agreement in 1944, the World Bank was designed to provide loans and grants for economic development (its formal name is actually the International Bank for Reconstruction and Development). Since its inception, it has lent $333 billion (adjusted 1996 figures) for developmental projects. Famous World Bank funded projects include the massive dam construction programmes in Egypt (the Nasser Dam) and in Zambia (the Kariba Dam). Money for the World Bank is provided by rich industrialized countries such as the USA and wealthy oil-exporting states such as Saudi Arabia.

world-systems theory: An approach to global change which stresses the importance of changing modes of production. Since the fifteenth century, a capitalist world economy has provided a powerful catalyst for social, economic and political change. The spatial organization of the world economy is divided into the core (major economies such as the USA, Japan and UK), the semi-periphery (nations such as Argentina, Australia, Indonesia, New Zealand and South Africa) and the periphery (the nations of the South). These categories are not considered static but are the result of a particular set of geographical and historical outcomes. World-systems analysts hope that by studying the 'big picture' they will be better able to predict whether it will be possible to change the unequal nature of the capitalist world economy. Most hope that a socialist (or at least more egalitarian) form of production might be possible. Other scholars are uncomfortable with the sweeping historical and geographical analyses of world-systems analysis.

World Trade Organization: The WTO was created in 1994 following a series of lengthy negotiations over the regulation of world trade labelled the Uruguay Round of talks. Its purpose, to function as a regulatory body, entails the responsibility of ensuring that the principles of free trade and fair competition are upheld in the world economy. Critics in the South complain that the WTO does not have sufficient power to prevent the major trading economies such as the USA and China from pushing an agenda which exposes fragile economies to unregulated competition and minimum controls on worker rights and environmental protection.

REFERENCES
AND FURTHER READING

Afshar, H. (ed.) 1998. *Women and Empowerment* Basingstoke: M acmillan.

Agnew, J. 1992. 'The US position in the world geopolitical order after the Cold War' *Professional Geographer* 44: 7–10.

Agnew, J. 1983. 'An excess of national exceptionalism: towards a new political geography of American foreign policy' *Political Geography Quarterly* 2: 151–66.

Agnew, J. 1994. 'The territorial trap: the geographical assumptions of international relations theory' *Review of International Political Economy* 1: 53–80.

Agnew, J. 1998. *Geopolitics* London: Routledge.

Agnew, J. 2002. *Geopolitics* London: Routledge (second edition).

Agnew, J. 2003. *Making Political Geography* London: Arnold.

Agnew, J. and S. Corbridge 1995. *Mastering Space* London: Routledge.

Agnew, J. and J. Sharp 2002. 'America, frontier nation: from abstract space to worldly place' in J. Agnew and J. Smith (eds.) *America Space/America Place* Edinburgh: Edinburgh University Press 79–107.

Agnew, J., K. Mitchell and G. O Tuathail (eds.) 2003. *A Companion to Political Geography* Oxford: Blackwell.

Ali, T. (ed.) 2000. *Masters of the Universe* London: Verso.

Allen, J. and D. Massey (eds.) 1995. *Geographical Worlds* Oxford: Oxford University Press.

Amnesty International. 2002. *Rights at Risk* London: Amnesty International.

Appy, C. 2000. 'Introduction' in C. Appy (ed.) *Cold War Constructions* Amherst: University of Massachusetts Press: 1–18.

Arab World Geographer. 2001. 'Forum' *Arab World Geographer* 4: 77–103.

Archibugi, D. 2002. 'Demos and cosmopolis' *New Left Review* 13: 24–40.

Arnold, G. 1993. *The End of the Third World* Basingstoke: Macmillan.

Athanasiou, T. 1997. *The Ecology of a Dying Planet* London: Secker and Warburg.

Ayoob, M. 1993. 'The new-old disorder in the Third World' *Global Governance* 1: 59–78.

Ayoob, M. 1995. *The Third World Security Predicament: State Making, Regional Conflict, and the International System* Boulder: Lynne Rienner.

Barber, B. 1996. *Jihad vs. McWorld* New York: Ballantine Books.

Barker, C. 1998. *Global Television* London: Routledge.

Barnett, T. 2003. 'The Pentagon's new map' *Esquire* March: 1–10.

Barton, J. 1997. *A Political Geography of Latin America* London: Routledge.

Baylis, J. and S. Smith (eds.) 2001. *The Globalization of World Politics* Oxford: Oxford University Press (second edition).

Beck, U. 1992. *The Risk Society* London: Sage.

Berg, E. 2003. 'Some unintended consequences of geopolitical reasoning in post-Soviet Estonia: texts and policy streams, maps and cartoons' *Geopolitics* 8: 101–20.

Berger, M. 1994. 'The end of the Third World' *Third World Quarterly* 15: 257–75.

Black, J. 2001. *The Politics of James Bond* New York: Praeger.

Block, F. 1996. *The Vampire State and Other Stories* New York: New Press.

Blum, W. 2002. *Rogue State* London: Zed.

Blunt, A. and J. Willis 2000. *Dissident Geographies* Harlow: Pearson Education.

Bodansky, Y. 2002. *Bin Laden: The Man who Declared War on America* (New York, Forum).

Booth, K. and T. Dunne 2002. *Worlds in Collision* Basingstoke: Palgrave.

Boutros-Ghali, B. 1992. *An Agenda for Peace* New York: United Nations Publication Unit.

Bowman, I. 1921. *The New World* New York: World Books.

Bradshaw, M. and A. Stenning (eds.) 2004. *East Central Europe and the former Soviet Union* Harlow: Pearson Education.

Breslin, S. 1996. *China in the 1980s*. Basingstoke: Macmillan.

Bretherton, C. 1996. 'Introduction: global politics in the 1990s' in C. Bretherton and G. Ponton (eds.) *Global Politics* Oxford: Blackwell 1–20.

Brown, C. 1997. *Understanding International Relations* London: Macmillan.

Buckley, M. and R. Fawn (eds.) 2003 *Global Responses to Terrorism* London: Routledge.

Bull, H. 1965. *The Control of the Arms Race* London: Methuen.

Bull, H. 1977. *The Anarchical Society* London: Macmillan.

Burchill, S. and A. Linklater (eds.) 1996. *Theories of International Relations* Basingstoke: Macmillan.

Burke, J. 2003. *Al Qaeda: Casting a Shadow of Terror* London: I. B. Tauris.

Bush, E. and L. Harvey. 1997. 'Joint implementation and the ultimate objective of the UN Frame-work Convention on Climate Change' *Global Environmental Change* 7: 265–86.

Byrne, D. 2004. *Human Rights* Harlow: Pearson Education.

Campbell, D. 1992. *Writing Security* Manchester: Manchester University Press.

Campbell, D. 1999. 'Contradictions of a lone superpower' in D. Slater and P. Taylor (eds.) *The American Century* Oxford: Blackwell: 222–40.

Carruthers, S. 1995. *Winning Hearts and Minds* Leicester: Leicester University Press.

Castaneda, J. 1994. *Utopia Unarmed* New York: Vintage.

Castells, M. 1996. *The Rise of the Network Society* Oxford: Blackwell.

Castles, S. and M. Miller 1993. *The Age of Migration* Basingstoke: Macmillan.

Castree, N. 2003. 'The geopolitics of nature' in J. Agnew, K. Mitchell and G. Toal (eds.) *A Companion to Political Geography* Oxford: Blackwell 423–39.

Chaturvedi, S. 1996. *Polar Regions: A Political Geography* Chichester: John Wiley.

Chaturvedi, S. 1998. 'Common security? Geopolitics, development, South Asia and the Indian Ocean' *Third World Quarterly* 19: 701–24.

Chomsky, N. 1991. *Deterring Democracy* London: Verso.

Chomsky, N. 2001. *9-11* New York: Seven Stories Press.

Cirincione, J. 2000. *Repairing the Regime: Preventing the Spread of WMD* London: Routledge.

Clarke, I. 1997. *Globalization and Fragmentation* Oxford: Oxford University Press.

Clifford, J. 1988. *The Predicament of Culture* Berkeley: University of California Press.

Cockburn, A., J. St Clair and A. Sekula 2000. *Five Days that Shook the World* London: Verso.

Cockburn, C. and D. Zarkov (eds.) 2001. *The Postwar Moment: Militaries, Masculinities and International Peace-Keeping* London: Lawrence and Wishart.

Cohn, C. 1987. 'Sex and death in the rational world of defense intellectuals' *Signs* 12: 687–718.

Cole, D. 2003. *Enemy Aliens* New York: New Press.

Collinson, S. 1996. *Shore to Shore: The Politics of Migration in Euro-Maghreb Relations* London: Royal Institute of International Affairs.

Commission on Global Governance. 1995. *Our Global Neighbourhood* Oxford: Oxford University Press.

Connell, J. and C. Gibson 2003. *Sound Tracks* London: Routledge.

Corbridge S., R. Martin and N. Thrift (eds.) 1994. *Money, Space and Power* Oxford: Blackwell.

Cosgrove, D. 1994. 'Contested global visions: one world, whole earth and the Apollo space photographs' *Annnals of the Association of the American Geographers* 84: 270–94.

Cotton, J. 2003. 'The second North Korea nuclear crisis' *Australian Journal of International Affairs* 57: 261–80.

Countryman, D. 1985. *The American Revolution* London: Penguin.

Cox, M. 2002. 'The new liberal empire' *Irish Studies of International Affairs* 12: 39–56.

Cox, R. 1981. 'Social forces, states and world orders' *Millennium* 10: 125–6.

Crang, M. 1998. *Cultural Geography* London: Routledge.

Cresswell, T. 1996. *In Place: Out of Place* Minneapolis: University of Minnesota Press.

Crockatt, R. 2003. *America Embattled: September 11th, Anti-Americanism and the Global Order* London: Routledge.

Dalby, S. 1990. *Creating the Second Cold War* London: Belhaven Press.

Dalby, S. 1991. 'Critical geopolitics: discourse, difference and dissent' *Environment and Planning D: Society and Space* 9: 261–83.

Dalby, S. 1993. 'The Kiwi disease: geopolitical discourse in Aotearoa, New Zealand and the South Pacific' *Political Geography* 12: 437–56.

Dalby, S. 2002. *Environmental Security* Minneapolis: University of Minnesota Press.

Dalby, S. 2003. 'Geopolitics, the Bush doctrine and the war on Iraq' *Arab World Geographer* 6: 7–18.

Danaher, K. and P. Burbach (eds.) 2000. *Globalise This!* New York: Common Courage Press.

Danchev, A. (ed.) 1995. *Fin de Siècle* London: Macmillan.

Davies, P. and P. Wells (eds.) 2002. *American Films and Politics from Reagan to Bush Jnr* Manchester: Manchester University Press.

Der Derian, J. 1992. *Anti-Diplomacy* Oxford: Blackwell.

Der Derian, J. 2001. *Virtuous War* Boulder: Westview.

Desai, V. and R. Imrie 1998. 'The new managerialism in local governance: North–South dimensions' *Third World Quarterly* 19: 635–50.

Desai, V. and R. Potter 2002. *The Companion to Development Studies* London: Arnold.

Dickins, P. 1992. *Global Shift* London: Paul Chapman Publishing.

Dijink, G. 1996. *National Identity and Geopolitical Visions* London: Routledge.

Dines, G. 1995. 'Towards a sociological analysis of cartoons' *Humor* 8: 237–55.

Dobson, A. 1990. *Green Political Thought* London: Routledge.

Dodds, K. 1996. 'The 1982 Falklands War and a critical geopolitical eye: Steve Bell and the If cartoons' *Political Geography* 15 (6/7): 571–92.

Dodds, K. 1997. *Geopolitics in Antarctica: Views from the Southern Oceanic Rim* Chichester: John Wiley.

Dodds, K. 1998. 'Political Geography I: the globalization of world politics' *Progress in Human Geography* 23: 595–606.

Dodds, K. 2000. 'Geopolitics, Patagonian Toothfish and living resource regulation in the Southern Ocean' *Third World Quarterly* 21: 229–46.

Dodds, K. 2003a. 'Licensed to stereotype: popular geopolitics, James Bond and the spectre of Balkanism' *Geopolitics* 8: 125–56.

Dodds, K. 2003b. 'Cold War geopolitics' in J. Agnew, K. Mitchell and G. O Tuathail (eds.) *A Companion to Political Geography* Oxford: Blackwell.

Dodds, K. and D. Atkinson (eds.) 2000. *Geopolitical Traditions* London: Routledge.

Dorrance, J. 1985. 'The South Pacific: emerging security issues and US policy' *Pacific Review* 4: 385–6.

Dougherty, B. 2002. 'Comic relief: using political cartoons in the classroom' *International Studies Perspectives* 3: 258–70.

Douglas, R., L. Harte and J. O'Hara (eds.) 1988. *Drawing Conclusions: A Cartoon History of Anglo-Irish Relations 1798–1998* Belfast: Blackstaff Press.

Doty, R. 1996. *Imperial Encounters* Minneapolis: University of Minnesota Press.

Doyle, T. 1998. 'Sustainable development and Agenda 21: the secular bible of global free markets and pluralist democracy' *Third World Quarterly* 19: 771–86.

Doyle, T. and D. McEachern 1998. *Environment and Politics* London: Routledge.

Elliot, L. 1998. *The Global Politics of the Environment* London: Routledge.

El-Nawawy, M. and A. Iskandar 2002. *Al Jazeera* Boulder: Westview Press.

Enloe, C. 1989. *Bananas, Bases and Beaches: Making Feminist Sense of International Politics* London: Pandora.

Enloe, C. 1993. *The Morning After: Sexual Politics at the End of the Cold War* Berkeley: University of California Press.

Escobar, A. 1995. *Encountering Development* Princeton: Princeton University Press.

Esposito, J. 2002. *What Everyone Needs to Know About Islam* Oxford: Oxford University Press.

Falk, R. 1995. *On Humane Governance* Oxford: Polity.

Falk R. 1997. 'Will globalization win out?' *International Affairs* 73: 123–36.

Falk, R. 1999. *Predatory Globalization* Oxford: Polity.

Fanon, F. 1961. *The Wretched of the Earth* London: MacGibbon and Kee.

Fanon, F. 1967. *Black Skins, White Masks* New York: Grove.

Ferguson, N. 2004. *Colossus* Harmondsworth: Penguin.

Flint, C. 2003. 'Terrorism and counter terrorism: geographical research questions and agendas' *Professional Geographer* 55: 162–9.

Foot, M. 1999. *Dr Strangelove I presume* London: Michael Joseph.

Frank, A. G. 1971. *Capitalism and Under-Development in Latin America* Harmondsworth: Penguin.

Freedman, L. and E. Karsh. 1993. *The Gulf War and the New World Order* Basingstoke: Macmillan.

French, P. 1998. *Liberty or Death* London: Abacus.

Friedman, T. 2002. *The Lexus and the Olive Tree* New York: Knopf.

Frost, M. 1996. *Ethics in International Relations* Cambridge: Cambridge University Press.

Frum, D. 2003. *The Right Man* London: Weidenfeld and Nicolson.

Frum, D. and R. Pearle 2003. *An End to Evil* New York: Random House.

Fukuyama, F. 'The end of history' 1989. *National Interest* 16: 1–8.

Fukuyama, F. 1992. *The End of History and the Last Man* New York: Free Press.

Furedi, F. 1994. *Colonial Wars and the Politics of the Third World* London: I. B. Tauris.

Galeano, E. 1973. *The Open Veins of Latin America* New York: Monthly Review Press.

Gamble, A. and A. Payne (eds.) 1996. *Regionalism and World Order* Basingstoke: Macmillan.

Gardner, G. 1994. *Nuclear Nonproliferation* Boulder: Lynne Rienner.

Gearty, C. and A. Tomkins (eds.) 1996. *Understanding Human Rights* London: Mansell.

Geopolitics (2003) '11th September and its aftermath: the geopolitics of terror' *Geopolitics* 8: 1–294.

Giddens, A. 1996. *Modernity and Self-Identity* Cambridge: Polity.

Gilpin, R. 1987. *The Political Economy of International Relations* Princeton: Princeton University Press.

Glassner, M. I. 1996. *Political Geography* Chichester: John Wiley.

Gould, S. 1998. *Questioning the Millennium* London: Vintage.

Gow, J. 1997. *Triumph of the Lack of Will: International Diplomacy and the Yugoslav War* London: C. Hurst.

Gowing, N. 1994. 'Real Time Television Coverage of Armed Conflict and Diplomatic Crises: Does it Pressure or Distort Foreign Policy Decisions?' Working Paper, John F. Kennedy School of Government, University of Harvard.

Graham, G. 1996. *Ethics and International Relations* Oxford: Blackwell.

Gramsci, A. 1971. *Prison Notebooks of Antonio Gramsci* New York: International Publishers.

Grant, C. 1995. 'Equity in international relations: a third world perspective' *International Affairs* 17: 561–587.

Gray, J. 2002. *Al Qaeda and What it Means to be Modern* London: Faber and Faber.

Gregg, R. 1998. *International Relations on Film* Boulder: Lynne Rienner.

Green, D. 1995. *Silent Revolution* London: Cassell.

Gunaratna, R. 2002. *Inside Al Qaeda* London: C. Hurst and Company.

Gupta, J. 1997. *The Climate Change Convention and Developing Countries* Dordrecht: Kluwer.

Halliday, F. 1989. *Cold War, Third World* London: Verso.

Halliday, F. 2001. *Two Hours that Shook the World* London: Saqi Books.

Hanlon, J. 1986. *Beggar Your Neighbours: Apartheid Power in Southern Africa* Oxford: James Curry.

Hanlon, J. 1996. *Peace Without Profit: How the IMF Blocks Rebuilding in Mozambique* Oxford: James Currey.

Hardt, M. and A. Negri 2000. *Empire* Cambridge, MA: Harvard University Press.

Harrison, P. and R. Palmer. 1986. *News Out of Africa* London: Hilary Shipman.

Harriss J. (ed.) 1995. *The Politics of Humanitarian Intervention* London: Pinter.

Haynes, J. 1996. *Third World Politics* Oxford: Blackwell.

Haynes, J. 2002. *Politics in the Developing World* Oxford: Blackwell.

Heffernan, M. 1998. *The Meaning of Europe* London: Arnold.

Held, D. 1995. *Democracy and the Global Order* Cambridge: Polity.

Held, D. 2004. 'Afterword' in D. Held (ed.) *A Globalising World? Culture, Economics and Politics* London: Routledge: 169–77.

Held, D. *et al.* (eds.) 1999. *Global Transformations* Oxford: Polity.

Hepple, L. W. 1986. 'The revival of geopolitics' *Political Geography Quarterly* 5: 21–36.

Hepple, L. W. 1992. 'Metaphor, geopolitical discourse and the military in South America' in T. Barnes and J. Duncan (eds.) *Writing Worlds* London: Routledge: 136–55.

Hepple, L. W. 2000. 'Géopolitiques de gauche: Yves Lacoste, Herodote and French radical politics' in K. Dodds and D. Atkinson (eds.) *Geopolitical Traditions* London: Routledge 268–301.

Herman, E. and N. Chomsky 1988. *Manufacturing Consent* London: Pantheon.

Herod, A., G. O Tuathail and S. Roberts (eds.) 1998. *An Unruly World?* London: Routledge.

Hersh, S. 1998. *The Dark Side of Camelot* London: HarperCollins.

Hertsgaard, M. 2002. *The Eagle's Shadow* London: Bloomsbury.

Higgott, R. 1997. 'De facto and de jure regionalism: the double discourse of regionalism in the Asia Pacific' *Global Society* 11: 165–85.

Higgot, R. and K. Nossal 1997. 'The international politics of liminality: relocating Australia in the Asia-Pacific' *Australian Journal of Political Science* 32: 169–86.

Hirst, P. and G. Thompson 1996. *Globalization in Question* Cambridge: Polity.

Hitchens, C. 2001. *The Trial of Henry Kissinger* London: Verso.

Hobsbawm, E. 1997. *On History* London: Weidenfeld and Nicolson.

Honig, J. and N. Both 1995. *Srebrenica* Harmondsworth: Penguin.

Hoogvelt, A. 1997. *Globalization and the Post-Colonial World* Basingstoke: Macmillan.

Hosle, V. 1992. 'The third world as a philosophical problem' *Social Research* 59: 227–62.

Huntington, S. 1993. 'The clash of civilisations' *Foreign Affairs* 72: 22–49.

Huntington, S. 1996. *The Clash of Civilizations and the Remaking of World Order* New York: Simon and Schuster.

Huque, M. 1997. 'Nuclear proliferation in South Asia' *International Studies* 34: 1–14.

Hurrell, A. 1995. 'International political theory and the global environment' in K. Booth and S. Smith (eds.) *International Relations Theory Today* Cambridge: Polity 129–53.

Hurrell, A. 1998. 'Security in Latin America' *International Affairs* 74: 529–46.

Hyndman, J. 'Beyond Either/Or: a feminist analysis of September 11th, *ACME: an international e-journal of critical geographies* 2: 1–13.

Ignatieff, M. 2000. *Virtual War* London: Chatto and Windus.

Ignatieff, M. 2003. *Empire Lite* London: Minerva.

Imber, M. and J. Vogler (eds.) 1995. *International Regimes of Global Environmental Change* London: Routledge.

Immerman, R. 1982. *The CIA in Guatemala* Austin: University of Texas Press.

Independent. 2003. 'The Niger connection: what we know, what we don't know and what we will never be told' 10 July 2003.

Jackson, R. 1990. *Quasi-States* Cambridge: Cambridge University Press.

Jeffords, S. 1994. *Hard Bodies: Hollywood Masculinity in the Reagan Era* New Brunswick, NJ: Rutgers University Press.

Johnston, R., D. Gregory and D. Smith (eds.) 2000. *The Dictionary of Human Geography* Oxford: Blackwell.

Kagan, R. 2002. 'Power and Paradise' *Policy Review* 113: 1–18.

Kagan, R. 2003. *Paradise and Power: America and Europe in the New World Order* New York: Monthly Press.

Kaldor, M. 1999. *New and Old Wars* Oxford: Polity.

Kaplan, R. 1994. 'The coming anarchy' *Atlantic Monthly* (Feb.): 44–76.

Kaplan, R. 2000. *The Coming Anarchy* New York: Vintage Books.

Kato, M. 1993. 'Nuclear globalism: traversing rockets, satellites and nuclear war' *Alternatives* 18: 339–60.

Kellner, D. 1992. *The Persian Gulf TV War* Boulder: Lynne Rienner.

Kellner, D. 1995. *Media, Culture and Society* London: Routledge.

Keohane, R. and J. Nye 1989. *Power and Interdependence* Cambridge, MA: Harvard University Press.

Kepel, G. 2002. *Jihad: The Trail of Political Islam* Cambridge: Cambridge University Press.

Khor, M. *Rethinking Globalization* London: Zed Books.

Klare, M. 2001. *Resource Wars* New York: Norton.

Kolko, G. 1997. *Vietnam: Anatomy of Peace* London: Routledge.

Kong, L. 1995. 'Popular music in geographical analysis' *Progress in Human Geography* 19: 183–98.

Krasner, S. 'Introduction' in S. Krasner (ed.) *International Regimes* Ithaca: Cornell University Press: 1–19.

Kristof, L. 1960. 'The origins and evolution of geopolitics' *Journal of Conflict Resolution* 4: 15–51.

Lacoste, Y. 1973. 'An illustration of geographical warfare: bombing of the dykes on the Red River, North Vietnam' *Antipode* 5: 1–13.

Lacoste, Y. 1976. *La Géographie, ça sert, d'abord, à faire la Guerre* Paris: Maspero.

Lane, D. 1996. *The Rise and Fall of State Socialism* Cambridge: Polity.

Lange, D. 1990. *Nuclear Free: the New Zealand Way* Auckland: Penguin.

Latin American Outlook (1995) 'Taking Mexico to market' Speech delivered by Carlos Heredia to the Latin American Bureau March 1995.

Lauterpacht, E. 1997. 'Sovereignty: myth or reality' *International Affairs* 73: 137–50.

Lewis, B. 2001. *What went Wrong? The Clash between Islam and Modernity in the Middle East* Oxford: Oxford University Press.

Lewis, B. 2004. *The Crisis of Islam* New York: Random House.

Leyshon, A., D. Matless and G. Revill 1995. 'The place of music' *Transactions of the Institute of British Geographers* 20: 423–33.

Liftin, K. 2003. 'Planetary geopolitics' in J. Agnew, K. Mitchell and G. Toal (eds.) *A Companion to Political Geography* Oxford: Blackwell 470–82.

Lifton, R. and R. Falk 1982. *Indefensible Weapons* New York: Basic Books.

Linklater, A. 1998. *The Transformation of Political Community* Cambridge: Polity.

Lipschultz, R. and K. Conca (eds.) 1993. *The State and Social Power in Global Environmental Politics* New York: Columbia University Press.

Love, J. 1980. 'Paul Prebisch and the origins of the doctrine of unequal exchange' *Latin American Research Review* 15: 45–72.

Luke, T. 2003. 'Postmodern geopolitics: the case of the 9.11 terrorist attacks' in J. Agnew, K. Mitchell and G. O Tuathail (eds.) *A Companion to Political Geography* Oxford: Blackwell: 219–35.

Mackinder, H. 1904. 'The geographical pivot of history' *Geographical Journal* 23: 421–44.

Malcolm, N. 1996. *Bosnia: A Short History* London: Macmillan.

Malcolm, N. 1998. *Kosovo: A Short History* London: Macmillan.

Mann, M. 2003. *Incoherent Empire* New York: Verso.

Massey, D. 1991. 'A global sense of place' *Marxism Today* (June): 25–6.

Mazarr, M. 1995. *North Korea and the Bomb* Basingstoke: Macmillan.

Mazower, M. 1998. *Dark Continent: Europe's Twentieth Century* Harmondsworth: Penguin.

McCannon, J. 1998. *Red Arctic* Oxford: Oxford University Press.

McFarlane, T. and I. Hay 2003. 'The battle for Seattle: protest, popular geopolitics, and the Australian newspaper' *Political Geography* 22: 211–32.

McGrew, A. 1998. 'The globalization debate: putting the advanced capitalist state in its place' *Global Society* 12: 299–322.

McGrew, A. 2004. 'Power shift: from national government to global governance?' in D. Held (ed.) *A Globalizing World? Culture, Economics and Politics* London: Routledge: 127–68.

Mermin, E. 1997. 'TV news and the United States intervention in Somalia' *Political Science Quarterly* 112: 385–403.

Middleton, N., P. O'Keefe and S. Mayo 1993. *Tears of the Crocodile* London: Pluto.

Miller, M. 1995. *The Third World in Global Environmental Politics* Milton Keynes: Open University Press.

Mittleman, J. (ed.) 1996. *Globalization: Critical Reflections* Boulder: Lynne Rienner.

Mittleman, J. 2000. *Globalization Syndrome* Princeton: Princeton University Press.

Monbiot, G. 2002. *The Age of Consent: A Manifesto for a New World Order* London: Flamingo.

Morgenthau, H. 1948. *Politics Among Nations* New York: Alfred Knopf.

Morley, D. and K. Robins. 1995. *Spaces of Identity* London: Routledge.

Mullerson, R. 1996. *Human Rights Diplomacy* London: Routledge.

Naisbitt, J. 1995. *Global Paradox* New York: Avon Books.

Newhouse, J. 1989. *The Nuclear Age* London: Michael Joseph.

Newkey-Burden, C. 2003. *Nuclear Paranoia* London: Pocket Essentials.

Nijman, J. 1992. *The Geopolitics of Power and Conflict* Chichester: John Wiley.

Nino, C. S. 1995. *Radical Evil on Trial* New Haven: Yale University Press.

Norton-Taylor, R. 1998. 'Comic strip *Animal Farm* used as cold war weapon' *Guardian* 17 March.

O'Byrne, D. 2003. *Human Rights* Harlow: Pearson Education.

Ohmae, K. 1990. *The Borderless World* London: Collins.

O'Loughlin, J. 1989. 'World power competition and local conflicts in the Third World' in R. Johnston and P. Taylor (eds.) *A World in Crisis* Oxford: Blackwell 289–332.

O'Loughlin, J. 1992. 'Ten scenarios for a new world order' *Professional Geographer* 44: 24–8.

O'Loughlin, J. (ed.) 1994. *The Dictionary of Geopolitics* Washington: Greenwood Press.

O'Loughlin, J. and R. Grant 1990. 'The political geography of presidential speeches 1946–1987' *Annals of the Association of American Political Geographers* 80: 504–30.

O Tuathail, G. 1994. 'Displacing geopolitics: writing on the maps of global politics' *Environment and Planning D: Society and Space* 12: 525–46.

O Tuathail, G. 1996. *Critical Geopolitics* London: Routledge.

O Tuathail, G. 1997. 'Emerging markets and other simulations: Mexico, the Chiapas revolt and geopolitical panopticon' *Ecumene* 4: 300–17.

O Tuathail, G. 1999. 'The ethnic cleansing of a "safe area": the fall of Srebrenica and the ethics of un-governability' in J. Proctor and D. Smith (eds.) *Geography and Ethics* London: Routledge: 120–31.

O Tuathail, G. 2000. 'The postmodern geopolitical condition: states, statecraft and security at the millennium' *Annals of the Association of American Geographers* 90: 166–78.

O Tuathail, G. and J. Agnew 1992. 'Geopolitics and foreign policy: practical geopolitical reasoning in American foreign policy' *Political Geography* 11: 190–204.

O Tuathail, G. and S. Dalby (eds.) 1998. *Rethinking Geopolitics* London: Routledge.

O Tuathail, G., S. Dalby and P. Routledge (eds.) 1998. *The Geopolitics Reader* London: Routledge.

O Tuathail, G. and T. Luke 1994. 'Present at the (dis)integration: deterritorialisation and reterritorialisation in the new wor(l)d order' *Annals of the Association of American Geographers* 84: 381–94.

Painter, J. 1995. *Politics, Geography and Political Geography* London: Arnold.

Parker, G. 1985. *Western Geopolitical Thought* London: Croom Helm.

Parkins, C. 1996. 'North–South relations and globalization after the Cold War' in C. Bretherton and G. Ponton (eds.) *Global Politics* Oxford: Blackwell 49–73.

Paterson, M. 1996. *Global Warming and Global Politics* London: Routledge.

Paterson, M. 2000. 'Car culture and global environmental politics' *Review of International Studies* 26: 253–70.

Paton Walsh, N. 2003. 'Hell on earth' *Guardian* 18 April.

Peterson, V. and A. Runyan 1993. *Global Gender Issues* Boulder: Westview.

Pettman, J. 1996. *Worlding Women* London: Routledge.

Phythian, M. 2000. *The Politics of Arms Sales since 1964* Manchester: Manchester University Press.

Piel, G. 1992. *Only One World* New York: United Nations.

Pilger, J. 1998. *Secret Agendas* London: Vintage.

Pinochet, A. 1984. *Geopolitica* Santiago de Chile: Editorial Andres Bello.

Pion-Berlin, D. 1989. *Ideology of State Terror* Boulder: Lynne Rienner.

Porter, G. and J. Brown. 1996. *Global Environmental Politics* Boulder: Westview Press.

Power, M. 2003. *Rethinking Development Geographies* London: Routledge.

Princern, T. and M. Finger. 1994. *Environmental NGOs in World Politics* London: Routledge.

Prins, G. (ed.) 1993. *Threats Without Enemies* London: Earthscan.

Proctor, J. and D. Smith (eds.) 2000. *Geography and Ethics* London: Routledge.

Prunier, G. 1995. *Rwanda: History of a Genocide 1959–1994* London: Hurst.

Pugh, M. 1996. 'Humanitarianism and peace-keeping' *Global Society* 10: 205–24.

Raban, J. 2003. 'The greatest gulf' *Guardian* 19 April.

Ramsbotham, O. and T. Woodhouse. 1996. *Humanitarian Intervention in Contemporary Conflict* Cambridge: Polity Press.

Redclift, M. 1987. *Sustainable Development* London: Methuen.

Reyntiens, F. 1995. *L'Afrique des Grands Lacs, Rwanda, Burundi 1988–1994* Paris: Les Afriques Karthala.

Rist, G. 1997. *History of Development* London: Zed.

Roberts, A. 1993. 'Humanitarian war: military intervention and human rights' *International Affairs* 69: 429–49.

Robertson, R. 1992. *Globalization: Social Theory and Global Culture* London: Sage.

Robinson F. 2001. 'On the Islamic background to the events of September 11th, Paper presented to the Yale Conference on Historical Perspectives on the September 11th Crisis 19 October 2001.

Robinson, P. 1999. 'The CNN effect: can the news media drive foreign policy?' *Review of International Studies* 25: 301–9.

Robinson, P. 2002. *The CNN Factor* Boulder: Westview Press.

Rogers, P. 2002. *Losing Control* London: Pluto Press.

Rose, G. 1993. *Feminism and Geography* Oxford: Polity.

Routledge, P. 1996. 'Critical geopolitics and terrains of resistance' *Political Geography* 15: 509–32.

Routledge, P. 1998. 'Going globile: spatiality, embodiment and media-tion in the Zapatista insurgency' in G. O Tuathail and S. Dalby (eds.) *Rethinking Geopolitics* London: Routledge: 240–60.

Routledge, P. 2003a. 'Anti-geopolitics' in J. Agnew, K. Mitchell and G. Toal (eds.) *A Companion to Political Geography* Oxford: Blackwell 236–48.

Routledge, P. 2003b. 'Convergence space: process geographies of grassroots globalisation networks' *Transactions of the Institute of British Geographers* 28: 333–49.

Said, E. 1978. *Orientalism* Harmondsworth: Penguin.

Said, E. 1981. *Representing Islam* Harmondsworth: Penguin.

Said, E. 1993. *Culture and Imperialism* London: Chatto and Windus.

Said, E. 1999. *Out of Place* Harmondsworth: Penguin.

Said, E. 2000. 'America's last taboo' *New Left Review* 6: 45–54.

Said, E. 2001. 'The clash of ignorance' *The Nation* 22 October: 9–12.

Sardar, Z. and M. Davies. 2002. *Why Do People Hate America?* London: Icon Books.

Schelling, R. 1992. 'Some economics of global warming' *American Economic Review* 82: 1–20.

Schlosser, E. 2003. *Reefer Madness and other Tales from the American Underground* Harmondsworth: Penguin.

Schmidt, B. 1998. *The Political Discourse of Anarchy: a Disciplinary History of International Relations* Albany: State University Press of New York.

Scholte, J. 1997. *Globalization: A Critical Introduction* Basingstoke: Macmillan.

Scholte, J. 2000. *Globalization* New York: St Martins Press.

Schulz, W. 2003. *Tainted Legacy* New York: Nation Books.

Seager, J. 1993. *Earth Follies* London: Earthscan.

Shafer, K. and A. Murphy 1998. 'The territorial strategies of IGOs: implications for environment and development' *Global Governance* 4: 257–74.

Shambroom, P. 2003. *Face to Face with the Bomb* Baltimore: Johns Hopkins University Press.

Sharp, J. 1993. 'Publishing American identity: popular geopolitics, myth and the Readers Digest' *Political Geography* 12: 491–503.

Sharp, J. 1996. 'Hegemony, popular culture and geopolitics: the Reader's Digest and the construction of danger' *Political Geography* 15: 557–70.

Sharp, J. 2000. *Condensing the Cold War: Reader's Digest and American Identity* Minneapolis: University of Minnesota Press.

Sharp, J., P. Routledge, C. Philo and R. Paddison (eds.) 2002. *Entanglements of Power: Geographies of Resistance and Domination* London: Routledge.

Shaw, M. 1996. *Civil Society and the Media in Global Crises* London: Pinter.

Shaw, T. 2001. *British Cinema and the Cold War* London: I. B. Tauris.

Shiva, V. 1993. 'The greening of global reach' in J. Childs and J. Cutler (eds.) *Global Visions: Beyond the New World Order* Boston: South End Press.

Shohat, E. and R. Stam. 1994. *Unthinking Ethnocentrism* London: Routledge.

Short, J. 1991. *Imagined Country* London: Routledge.

Sidaway, J. 1998. 'What's in a Gulf? From the "arc of crisis" to the Gulf War' in G. O Tuathail and S. Dalby (eds.) *Rethinking Geopolitics* London: Routledge: 224–40.

Sidaway, J. 2001. *Imagined Regional Communities* London: Routledge.

Sidaway, J. and D. Simon 1993. 'Geopolitical transition and state formation: the changing political geographies of Angola, Mozambique and Namibia' *Journal of Southern African Studies* 19: 6–28.

Simai, M. 1997. 'The changing state system and the future of global governance' *Global Society* 11: 141–64.

Simms, B. 2001. *Unfinest Hour: Britain and the Destruction of Bosnia* Harmondsworth: Penguin.

Simon, D. and K. Dodds (eds.) 1998. 'Rethinking geographies of North–South development' *Third World Quarterly* 19: 593–791.

Simon, D., W. van Spengen, C. Dixon and A. Narman (eds.) 1995. *Structurally Adjusted Africa* London: Pluto.

Simpson, G. 2003. *Great Powers and Outlaw States* Cambridge: Cambridge University Press.

Singer, P. 2004. *The President of Good and Evil* London: Granta.

Singham, A. and S. Hune. 1986. *Non-Alignment in the Age of Alignment* London: Zed.

Slater, R., B. Schultz and D. Dorr (eds.) 1993. *Global Transformation and the Third World* Boulder: Lynne Rienner.

Smith, A. 1995. *Nations and Nationalism in a Global Era* Cambridge: Cambridge University Press.

Smith, D. 2000. 'Moral progress in human geography: transcending the place of good fortune' *Progress in Human Geography* 24: 1–18.

Smith, N. 2001. 'Scales of terror and the resort to geography: September 11, October 7' *Environment and Planning D: Society and Space* 19: 631–7.

Smith, N. 2002. 'Global Seattle' *Environment and Planning D: Society and Space* 18: 1–5.

Smith, P. 1997. *Millennial Dreams* London: Verso.

Smith, S. 1994. 'Soundscape' *Area* 26: 232–40.

Smith, S. 1997. 'Beyond geography's visible worlds: a cultural politics of music' *Progress in Human Geography* 21: 502–29.

South Commission 1990. *The Challenge to the South* Oxford: Oxford University Press.

Sparke, M. 1998. 'Outside insides patriotism' in G. O Tuathail and S. Dalby (eds.) *Rethinking Geopolitics* London: Routledge: 198–223.

Spybey, T. 1996. *Globalization and World Society* Oxford: Polity.

St Clair, J. 1999. 'Seattle diary: it's a gas, gas, gas' *New Left Review* 238: 81–96.

Steger, M. 2003. *Globalisation: A Very Short Introduction* Oxford: Oxford University Press.

Stiglitz, J. 2002 *Globalisation and its Discontents* Harmondsworth: Penguin.

Stoett, P. 1994. 'The environment enlightenment: security analysis meets ecology' *Coexistence* 31: 127–47.

Stokke, O. and D. Vidas (eds.) 1996. *Governing the Antarctic* Cambridge: Cambridge University Press.

Taylor, D. 2003. *Orwell: The Life* London: Chatto and Windus.

Taylor, P. 1997. *Global Communications, International Affairs and the Media Since 1945* London: Routledge.

Taylor, P. J. 1993. *Political Geography* Harlow: Addison Wesley Longman.

Taylor, P. J. 1996. *The Way the Modern World Works* Chichester: John Wiley.

Taylor, P. J. and C. Flint. 2000. *Political Geography* Harlow: Pearson Education.

Thomas, C. 1987. *In Search of Security: The Third World in International Relations* Brighton: Harvester.

Thomas, C. and P. Wilkins (eds.) 1997. *Globalization in the South* Basingstoke: Macmillan.

Thompson, E. 1994. *The Age of Extremes* London: Verso.

Thrift, N. 1992. 'Muddling through: world order and globalization' *Professional Geographer* 44: 3–7.

Thrift, N. 1995. 'A hyperactive world' in R. Johnston, P. Taylor and M. Watts (eds.) *Geographies of Global Change* Oxford: Blackwell: 18–35.

Townsend, C. 2002. *Terrorism: A Very Short Introduction* Oxford: Oxford University Press.

United Nations. 1996. *The United Nations and Somalia 1992–1996* New York: United Nations Publications.

United Nations. 1998. *UN Human Development Report* New York: Oxford University Press.

United Nations. 2003. *UN Human Development Report* New York: Oxford University Press.

Urry, J. 2002. 'The global complexities of September 11th' *Theory, Culture and Society* 19: 57–70.

Vattimo, G. 1993. *The Transparent Society* Cambridge: Verso.

Vincent, J. 1974. *Non-Intervention and International Order* Princeton: Princeton University Press.

Virilio, P. 1986. *Speed and Politics* New York: Semiotext.

Virilio, P. 1989. *War and Cinema* London: Verso.

Virilio, P. 1997. *Open Sky* London: Verso.

Vogler, J. 1995. *Global Commons* Chichester: John Wiley.

Vogler, J. 1999. *Global Commons.* Chichester: John Wiley.

Vogler, J. and M. Imber (eds.) 1995. *The Environment and International Relations* London: Routledge.

Wainwright, J., S. Prudham and J. Glassman 2000. 'The battle in Seattle: microgeographies of resistance and the challenge of building alternative futures' *Environment and Planning D: Society and Space* 18: 5–13.

Walker, R. 1993. *Inside/Outside* Cambridge: Cambridge University Press.

Walker, R. 1998. *One World, Many Worlds: Struggles for a Just World Peace* Boulder: Lynne Rienner and Zed.

Walker, W. 1998. 'International nuclear relations after the Indian and Pakistani test explosions' *International Affairs* 74: 505–28.

Wallerstein, I. 1980. *The Modern World System II* New York: Academic Press.

Wallerstein, I. 2003. *The Decline of American Power* New York: New Press.

Waltz, K. 1979. *Theory of International Politics* Reading MA: Addison Wesley.

Waltz, K. 1981. *The Spread of Nuclear Weapons* London: Adelphi Paper Number 171.

Ward, J. 1997. *Latin America* London: Routledge.

Waters, M. 1995. *Globalization* London: Routledge.

Waters, M. 2001. *Globalization* London: Routledge, second edition.

Weiss, T. 1994. 'UN responses in the former Yugoslavia: moral and operational choices' *Ethics and International Affairs* 8: 1–22.

Weiss, T. and L. Collins 1996. *Humanitarian Intervention in the Post-Cold War Era* Boulder: Lynne Rienner.

Wellman, B. and C. Haythornhwaite (eds.) *The Internet in Everyday Life* Oxford: Blackwell.

Whitaker, J. 1997. *The United Nations* London: Routledge.

White, B., R. Little and M. Smith (eds.) 1997. *Issues in World Politics* Basingstoke: Macmillan.

Whitfield, S. 1991. *The Culture of the Cold War* Baltimore: Johns Hopkins University Press.

Willetts, P. 1978. *The Non-Aligned Movement* London: Pinter.

Williams, M. 1997. 'The Group of 77 and global environmental politics' *Global Environmental Change* 7: 295–8.

Woodward, S. 1995. *Balkan Tragedy* Washington DC: Brookings Institution.

Yair, G. 1995. 'Unite-unite Europe: the political and cultural structures of Europe in the Eurovision Song Contest' *Social Networks* 17: 147–61.

Young, R. 2003. *Postcolonialism: A Very Short Introduction* Oxford: Oxford University Press.

Younge, G. 2003. 'The US, race and war' *Guardian* 11 August 2003.

INDEX

T - #0025 - 161024 - C0 - 234/172/15 - PB - 9780273686095 - Gloss Lamination